MW00826455

MAKING TRACKS

To IAN
from Dad

Best Wishes
Sue Claassen
Ed Claassen

Making Tracks

C.L. Best and the Caterpillar Tractor Co.

Ed and Sue Claessen

ISBN: 978-1-59298-427-5
Library of Congress Control Number: 2011934463

BEAVER'S POND PRESS

Printed in the United States of America
First Printing: 2011
Book design by Ryan Scheife, Mayfly Design
15 14 13 12 11 5 4 3 2 1

Beaver's Pond Press, Inc.
7104 Ohms Lane, Suite 101, Edina, MN 55439-2129
(952) 829-8818 • www.BeaversPondPress.com
To order, visit www.BeaversPondBooks.com
or call (800) 901-3480. Reseller discounts available.

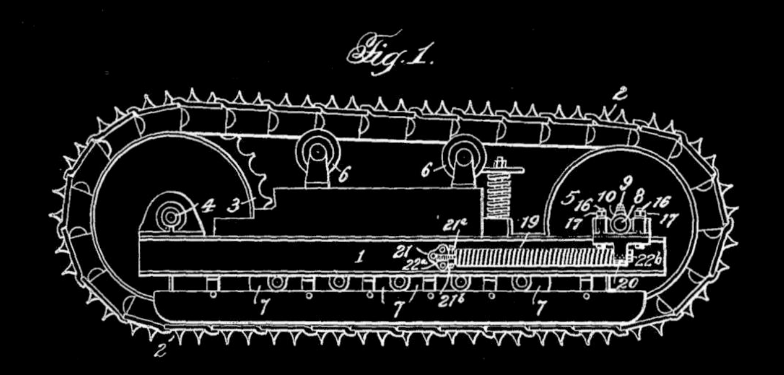

CONTENTS

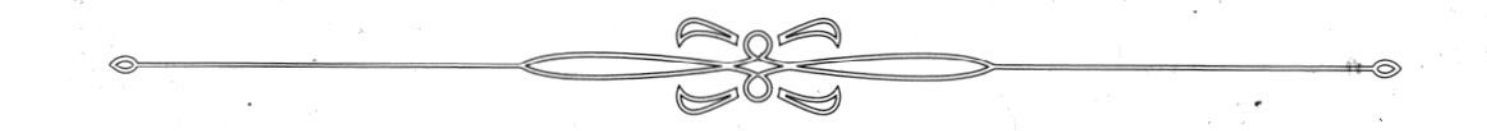

This book is dedicated to the Best family, both past and present.

To Dan G. Best, even though you are no longer here to read the completed book, we heeded your admonishment well: "Make sure you can prove it."

To Dan G. Best II, without your friendship, encouragement, and support, this book would have stayed in the research stage limbo.

And finally to C. L. Best, with the nudges you gave us to keep going and the feeling that now was indeed the time for this story to be told, we believe we did follow your wisdom: "If you haven't found the answer, you haven't worked hard enough."

We did and we have.

ACKNOWLEDGEMENTS

Many individuals are involved with any project six years in the making, some to a greater extent and some to a lesser extent, but all equally important to the completed project. So it was with *Making Tracks*.

The authors have no words to express just how crucial the support of the Best family was to this project. Please know that this book represents our best work for the current and future generations of the family. Much of the early Daniel Best history was already recorded by Daniel's great-grandsons, Terry and Brent Galloway, in the 1960s. We appreciate their previous work and being allowed access to their unpublished research. Frank Best, direct descendant of the Bests who didn't immigrate to the West Coast, provided additional "Uncle Daniel" family history, documents, and photos. John Weaver, C. L. Best's great-grandson, permitted us to reprint an original drawing made for his mother, Brenda, by her grandfather C. L. Best. Thank you, Best family.

We visited numerous institutions during our research for this book. Our heartfelt thanks go to John Skarstad, university archivist, and the staff at Special Collections, University of California Library, Davis; Kimberly Bray, curator of Archival Collections at the Haggin Museum in Stockton, California; and Mary Lee Barr and all the reference librarians and volunteers at the main branch of the San Leandro Public Library in San Leandro, California. We also thank the staffs at the California History Room and California State Library in Sacramento, California, and at the National Archives and Records Administration in San Bruno, California.

Special mention and thanks go to Jim Christianson, Dennis Munson, and Dan Rydland for sharing their technical knowledge and insights, and our thanks also go to Clarence Goodburn.

To Margy, Mel, and Jon Millerbernd, how do you thank people who opened their home and their hearts to our dog, Annie? If it weren't for the three of you, we would never have been comfortable being gone for the amount of time needed to complete our research.

And our final thanks go to the people who read our manuscript at its various stages of completion. All comments and suggestions were appreciated. So here is *Making Tracks*. We hope you enjoy this story that went untold for sixty years.

—Ed & Sue Claessen

FOREWORD

I met Ed Claessen in 1994. During that first visit to my ranch, he spoke at length about how impressed he was with my grandfather C. L. Best and the things that C. L. had accomplished with Best tractors and Caterpillar. After our discussion, we stood and shook hands, and as he prepared to take his leave, I invited him to get in touch with me when he was in California again. As I watched him get into his car and drive away, I thought to myself, "Who is this man from Minnesota, and why is he so enamored with MY grandfather?" When I was growing up, I always knew we were connected to the Caterpillar Tractor Co., but it wasn't part of our everyday lives. Back in 1929, my grandmother Pearl Best separated from C. L. and moved to Woodland with her daughter Betty and her son Dan, my father. The divorce became final in 1930. Grandfather remarried and lived at Mission San Jose for many years. When he died in 1951, I was about nine years old.

Ed came back to California the next year. When he called, I invited him back to the ranch. I was intrigued with his interest in C. L. and wanted to find out more. And Ed was up to the task. He gave me the background on the Best Sixty tractor that was on display by my office and why Grandfather was such an innovator. That was when I suggested that he should meet my father, Dan Best. Dad and Ed hit it off right away. I think it was because Ed could talk knowledgeably about many of the tractors that Dad used during the early years on his ranch. Dad and Ed, along with my mother, Bernice, had an enjoyable visit, and Ed was again invited back the next time he came West.

Over the following years, Ed and I had the opportunity to visit numerous times. As he shared more and more information about Grandfather, I began to realize just how important Grandfather's role was in the development of the modern tracklaying tractor. In 2003, during another California visit, I mentioned to Ed that someone should put all of the information he had acquired over the years into writing so it could be shared with the public. Written history and even C. L. himself downplayed the importance of his contributions to the industry. After we toyed with the idea of writing a book for a few years, the project came together when in 2006 Ed's wife, Sue, signed on as our writer. The intensive research and even more intense discussions continued until 2010 when we had a completed manuscript. The Best side of the story would finally see the light of day.

No more understatement, only the truth. I am proud of what we have created. I am only sorry that both my father and mother aren't here to see the completed project.

—Dan G. Best II

C. L. Best

PREFACE

Every company can trace its beginning to an idea, a concept, a dream. In the ensuing years of growth and prosperity, the early story of who, what, where, and when sometimes gets misplaced. When you're dealing with men who are secure with who they are and with what they can do, the need for personal recognition pales. They are satisfied to let their achievements speak through the years. But should they have been? When a man has patent designs that continue to be used in products for nearly one hundred years and are now considered the industry standard, shouldn't he deserve recognition? When the company he was instrumental in founding in 1925 has $48 billion in sales in 2008 and employs over one hundred thousand people worldwide, shouldn't his name and the names of his inner circle of advisors be known? And what if this company is an international giant with products recognized around the globe? Shouldn't the early journey leading to that success be acknowledged?

Ask people what they know about Caterpillar Inc., and most will think of huge yellow machines moving and sculpting the earth. Depending on where they live, some will think of machines preparing the soil and planting crops; some will picture forests being harvested; some will recall huge trucks moving vast amounts of ore from open-pit mines; some will think of the engines that power the trucks that move a nation's goods; some will think of the power units that keep the hospitals functioning when the power grid fails; and a few will even think of the Caterpillar name on clothing and boots. Ask about the beginnings of the Caterpillar Tractor Co., and a few will have heard about Benjamin Holt and perhaps Daniel Best. But who was C. L. Best? What could this man have accomplished that he should be remembered and revered for?

Clarence Leo Best, a son of Daniel Best, was born in 1878, when agricultural mechanization was only beginning. Man power would give way to horse power, horse power to steam power, steam power to internal combustion power, and internal combustion power to Diesel power. Many forward-thinking men proposed machines to ease the burden of the producer and also to increase profitability, while others refined existing patents to also reach those goals.

Born to Daniel and Meta Best in Albany, Oregon, C. L. Best was in the perfect position to be molded into the man whose ideas would change the world. Best would become the first chairman of the board of the Caterpillar Tractor Co. and remain so until his death twenty-six years later. To understand the mechanical and business genius for which C. L. Best should be known, the story begins on March 28, 1838, in Tuscarawas County, Ohio, when his father, Daniel Isaacs Best, was born to John George and Rebecca Best. One year later, John George moved his family to Missouri, where they stayed for about nine years. The Bests settled on a farm and "put up a mill on Bourbon Creek and sawed the lumber used in building the houses of the pioneers of that locality, and cherry lumber used in building coffins."[1] The sawmill consisted of moving parts and spinning gears as the saw blade sliced through the log—just the sort of thing to inspire Daniel's young, inquisitive mind. In 1847, his family moved near Vincennes, Iowa, on the Des Moines River. John George Best, his family now totaling eighteen, continued farming, but the Pacific Northwest would issue a strong call to some of his sons.

---

1. Terry R. Galloway and Brent D. Galloway, "Daniel Best Biography," *Engineers and Engines Magazine 14*, no. 5 (1969): 1.

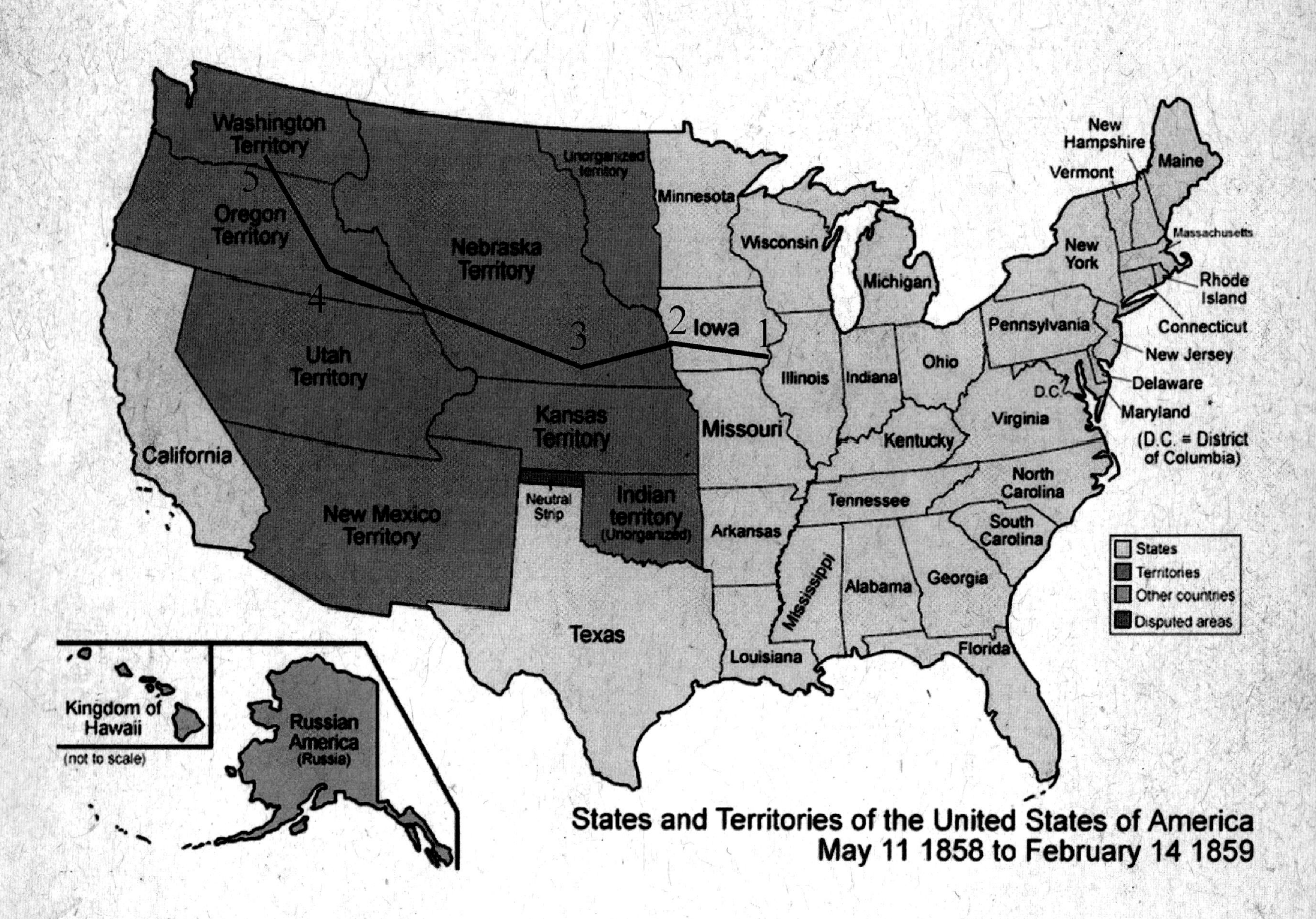

Daniel Best's travels 1859. Starting in Vincennes, Iowa (1), he traveled across the state to Council Bluffs (2). The next major stop on the Oregon Trail was Fort Kearny, Nebraska Territory (3). At Fort Hall, Oregon Territory (4), the original group split up. Best ended his journey at Fort Walla Walla, Washington Territory (5).

Lure of the West

Samuel Best, ten years older than his brother Daniel, was inspired by the discovery of gold in California in 1849 and left Iowa in 1852 for the riches of the Pacific Northwest. He was a miner in Althouse Creek, Oregon, and later worked on a farm at Waldo, Oregon. At the age of twenty-one, Daniel was more than ready to leave the farm and follow the Oregon Trail west. Despite his father's refusal to allow him to leave, Daniel was determined. In 1859, he joined a wagon train bound for Fort Walla Walla, Washington, and worked as a stock tender and sharp shooter. Wagon trains from Iowa journeyed to Council Bluffs, Iowa, which was one of the three major eastern starting points of the famous Oregon Trail. All three starting points, Independence and St. Joseph, Missouri, and Council Bluffs, joined at Fort Kearny on the Platte River in Nebraska.

The train consisted of about thirty wagons and around one hundred men, women, and children. Along the way, Daniel put his shooting skills to good use. He provided prairie chickens, jack rabbits, and other small game for the members of the wagon train. There were a number of skirmishes with the natives of the prairie: most involved livestock—horses, oxen, and cattle. Sometimes bargains were made; sometimes fighting broke out. Daniel G. Best, the grandson of Daniel Best, remembers this story about his grandfather:

> *We were at the duck club, sitting around talking: the Old Man [Daniel], Dad [C. L.], and me. Dad got the Old Man talking about the trip out to Oregon and the Indian troubles they had. Dad asked if he had to shoot any Indians. The Old Man said, "Well, one time there was an Indian in the bushes. I shot into the wiggling bushes and they stopped wiggling. I don't know—I might have winged him." You know, I always wondered about that story. The Old Man was a good shot; he hit*

whatever he aimed at. I figured he just didn't want to have to tell a kid like me exactly what had happened out there.[1]

The wagon train continued into Idaho. When they arrived at a nearly deserted Fort Hall, lack of supplies caused a change in plans. As stated by Daniel, "We had a conference. There wasn't enough food for all of us to go on with. Myself with thirteen companions, each with only a pocketful of biscuits, our guns, and ammunition, set out over five hundred miles of only the Lord-knew-what towards Fort Walla Walla."[2] This smaller party, with Daniel as one of the leaders, faced a wide variety of trials, but did arrive safely at Walla Walla.

For the next eight years, Daniel's interests revolved around lumbering, farming, and mining in Washington and Oregon. When faced with financial ruin in one industry, he used his natural abilities to find success in another. Farming and lumber milling provided him a stake to try gold mining; gold mining provided the stake to build a sawmill. In 1868, while operating a lumber mill near Olympia, Washington, a life-changing event occurred. An accident involving the mill's saw blade robbed him of his first three fingers, up to the second joint, on his left hand. Daniel was known to comment that if he hadn't lost his fingers, he wouldn't have used his head.[3]

Daniel's brother Henry and his wife, Lavina, had settled in Sutter County, California, near the towns of Yuba City and Marysville in 1862. With Henry and Lavina having arrived with three horses and all of their possessions in one wagon, Henry sold one horse to raise funds for survival until he could find employment. By 1864, Henry and Lavina were farming seventy rented acres of bottom-lands. The crops, mostly grains, were good, and the Bests netted $1,200 that year, which was used to purchase 160 acres of land. Over time, Henry brought his land holdings to about two thousand acres in the Sutter County area.[4]

1. Personal interview with Daniel G. Best, grandson of Daniel Best, April 6, 2006.
2. Terry R. Galloway and Brent D. Galloway, "Daniel Best Biography," *Engineers & Engines Magazine 14*, no. 5 (January–February 1969): 3.
3. Ibid., 4.
4. Peter J. Delay, *History of Yuba and Sutter Counties* (Los Angeles: Historic Records Company, 1924), 552.

Best's Improved Grain Separator patented April 25, 1871. San Leandro Public Library Historical Photo Collection #01585.

In 1869, Daniel was persuaded by his brother Henry to leave Washington and spend some time on Henry's ranch near Yuba City. Henry and his brothers, Samuel and Zachariah, raised grain. After threshing the grain heads from the stalks, the Bests and their Sutter County neighbors needed to haul their grain to town to have it cleaned and separated at a charge of $3.00 per ton. It was while staying with Henry that Daniel took in this arrangement and worked toward a profitable solution. If the cleaning and separating could be done on the farm, the savings in both time and money would be substantial. During the winter of 1869–70, Daniel built three separators, which were tested during the following harvest season. The portable separators were successful.[5] Daniel applied for a patent on this machine, and the patent was granted on April 25, 1871. This first patent

5. Terry R. Galloway and Brent D. Galloway, "Daniel Best Biography," *Engineers & Engines Magazine 14*, no. 5 (January–February 1969): 5.

Daniel Best wedding photo. 1872. San Leandro Public Library Historical Photograph Collection #001584.

Meta Johanna Steinkamp Best. San Leandro Public Library Historical Photograph Collection #01101.

was followed by forty-two additional patents that he was granted during his lifetime.

To gain capital to start production, Daniel sold half-interest in his separator to L. D. Brown for $5,000. The "Best and Brown's Unrivaled Seed Separator" was launched. It was advertised as having the capacity to separate thirty to sixty tons of grain per day. When the new machine made an appearance at the 1871 California State Fair, it was awarded first prize. Demand for the separators necessitated a formal shop be established for production. This was located in nearby Marysville, California.

Though building and operating the portable grain cleaners took up a good amount of Daniel's time, things were progressing on a personal level as well. On her parents' ranch in Yuba County, California, young Meta Steinkamp, daughter of John and Elizabeth Steinkamp, met Daniel Best, who at the time was working at his brother Henry's ranch. On August 29, 1872, Daniel and

Meta were married. They purchased a farm in the Marysville area. Daniel and his brothers were providing grain-cleaning services to the area farmers (using the separator based on Daniel's first patent) and were making a good profit. "[M]y brothers and myself ran those machines, cleaning from sixty to seventy tons a day with each one, for which we received $3.00 per ton. This meant an income of $180 to $210 per machine per day."[6]

Daniel and Meta moved to Auburn, Oregon, during the winter of 1874–75. Daniel was busy with mining interests. As shared by Meta in letters to her sister-in-law, Matilda Jane, he worked a number of claims and had other miners working for him. But unexpected problems arose, and he suffered near financial ruin. Not one to quit, Daniel regrouped and paid his debts.[7] He went from farming to mining to logging while still working on inventing. This chain of events was often repeated in various order, but the one constant throughout those early years was inventing.

The Best family spent the next years in the Willamette Valley. C. L. Best made his appearance on April 21, 1878. Later that same year, the Portland Board of Trade issued a requirement that prior to sale and shipment, grain had to be cleaned. Best, with his new partner Sam Althouse, set up shop in Albany to manufacture grain cleaners and separators. Sales were brisk. By 1879, a branch was opened in the Oakland, California, area. Daniel stated that "all during the winter we'd manufacture our implements and machinery up north; in the summer I'd go down to Oakland and act the part of a salesman and assembly expert."[8]

As his separator business was continuing in both Oregon and California, Daniel was patenting improvements to his separator/cleaner. Also during this time, Best worked as a pattern maker for Nathaniel Slate at the Cherry Iron Works in Albany, Oregon. According to Slate, "Dan Best was in his prime here, as I recall him. His boy Leo came out to the ranch with him one day while the combine was being built. We took a header and a separator and combined them."[9] That early attempt at a "combine" proved

Improved Hand Separator. Best Manufacturing Co. Circa 1893. San Leandro Public Library Historical Photograph Collection #01592.

unwieldy as it required too many horses to pull it to be practical. Slate put the idea aside.

Not content to manufacture only stationary separator/cleaners, Daniel Best set his mind to the problem of a traveling harvester that was both practical and efficient in operation. Working out of his assembly plant located at Third and Washington Streets in Oakland, Best applied for his seventh patent in 1885. This patent also concerned the cleaner. Running out of space in Oakland was later noted by his son, C. L. Best: "He had an assembly plant in Oakland, but the Police Department would not permit him to put these grain cleaners out on the streets, so he bought the San Leandro Plow Works."[10] The plow works was at that time owned by a local business consortium comprised of

6. "Happenings in San Leandro," *Oakland Tribune* (January 10, 1919): 8.

7. Ibid.

8. Ibid.

9. Ibid.

10. Letter from C. L. Best to L. B. Neumiller (August 27, 1943), F. Hal Higgins Collection, Special Collections, University of California Library, Davis.

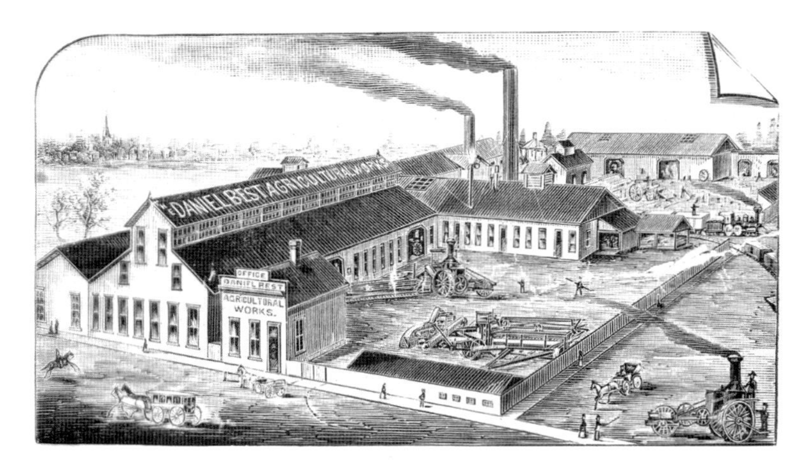

Daniel Best Agricultural Works. San Leandro, California. 1891 company catalog. San Leandro Public Library Historical Photograph Collection #01584.

Jacob Price, Socrates Huff, L. L. Morehouse, and Charles Gray. For the consideration of $15,000—to be paid with future profits—Daniel Best became the owner of the plow works, which consisted of the plow shop, planing mill, machine shop, foundry, and paint shop all located on a lot in block 50 on Davis Street. An article in the September 25, 1886, *Pacific Rural Press* commented on the Best Agricultural Works production facility:

> *The factory is considered one of the best equipped for work in the state. It has a splendid Corliss engine and boiler, shafting, saws, planers, a steam-driven hammer, a foundry, emery wheels, grinding apparatus for making plows, and much other necessary tools and machinery.*[11]

Daniel moved his Oakland operation the 8.25 miles to San Leandro. These two sites were connected by their proximity to the Southern Pacific Railroad, making it easier to move both production machinery and inventory.

The newly founded Daniel Best Agricultural Works continued to produce the Price Hay Press and Driver Harvester and added the Best Grain Cleaner. With business booming, it was necessary to add a 40'×80' grinding shop and drying room.[12] The horsepowered Best Combined Harvester made its commercial appearance in 1885. As told by Daniel Best in an interview in the *Oakland Tribune* of January 1919:

> *Throughout 1885, I continued to build cleaners, but at the same time developed the Best Combined Harvester, and in the fall of '86, sold six of these machines with from 12 to 20 foot cuts. In '87 I built sixty of these machines; in '88 one hundred and fifty.*[13]

Making a firm commitment to the San Leandro area, Daniel Best sold his business and personal interests in Oregon and moved his family southward in 1886. Located on the corner of Clarke Street and West Estudillo, the house originally built by Joseph DeMont (who was on the board of directors for the San Leandro Plow Company) allowed Daniel and Meta Best, along with their children, to live within easy walking distance of the Best Agricultural Works.

During the next two years, Daniel Best continued to produce and improve his harvesters and cleaners. In 1887, he applied for and was granted patents that combined a header to a threshing machine, which used a Best grain cleaner, to make a combined harvester and also a blast governor for his grain cleaners. This governor was a marked improvement. Regardless of the travel speed of the horses pulling the combine, it controlled the amount of wind passing through the cleaner, allowing more grain to be cleaned and not blown over and lost. This feature increased profitability—a fact that was not lost on the western farmers.

Daniel was concerned with customer satisfaction. A satisfied customer passed that information to others and became a repeat buyer. All models were thoroughly tested before delivery; the customer was given extensive instruction on its operation, and a generous free trial period was allowed before payment was due. Daniel wrote to each of his customers and received replies from many. On July 2, 1887, B. F. Walton of Yuba City, California, wrote:

11. "Daniel Best Agricultural Works," *Pacific Rural Press* (September 25, 1886): 4.

12. Harry Shaffer, *A Garden Grows in Eden: The Centennial Story of San Leandro* (San Leandro, CA: San Leandro Historical Centennial Committee, 1972): 50.

13. "Happenings in San Leandro," *Oakland Tribune* (January 10, 1919): 8.

> *Dear Sir: I have just put my Harvester away in good order for the season, after a very successful run of thirty days, cutting a little over 900 acres of grain! The machine has worked to a charm, and I am satisfied. Enclosed please find a check for $1700, which I enclose with pleasure, feeling I have value received for every cent of it.*[14]

Even though Best refined his harvester, the amount of horses or mules required to provide power for the machine was a major problem. Daniel's solution to that problem would begin with a visitor from Oregon.

(Above) *Train load of Best Combined Harvesters bound for American Falls, Idaho. Southern Pacific Railway.* San Leandro Public Library Historical Photograph Collection #01579. (Below) *California State Agricultural Society award to Daniel Best for Best Grain Cleaning Attachment for Combined Harvester.* 1887. Original medallion courtesy of Dan G. Best II.

14. Terry R. Galloway and Brent D. Galloway, "Daniel Best Biography," *Engineers & Engines Magazine 14*, no. 6 (March–April 1969): 3.

Daniel Best's Improved New Model

COMBINED HARVESTER.

1893.

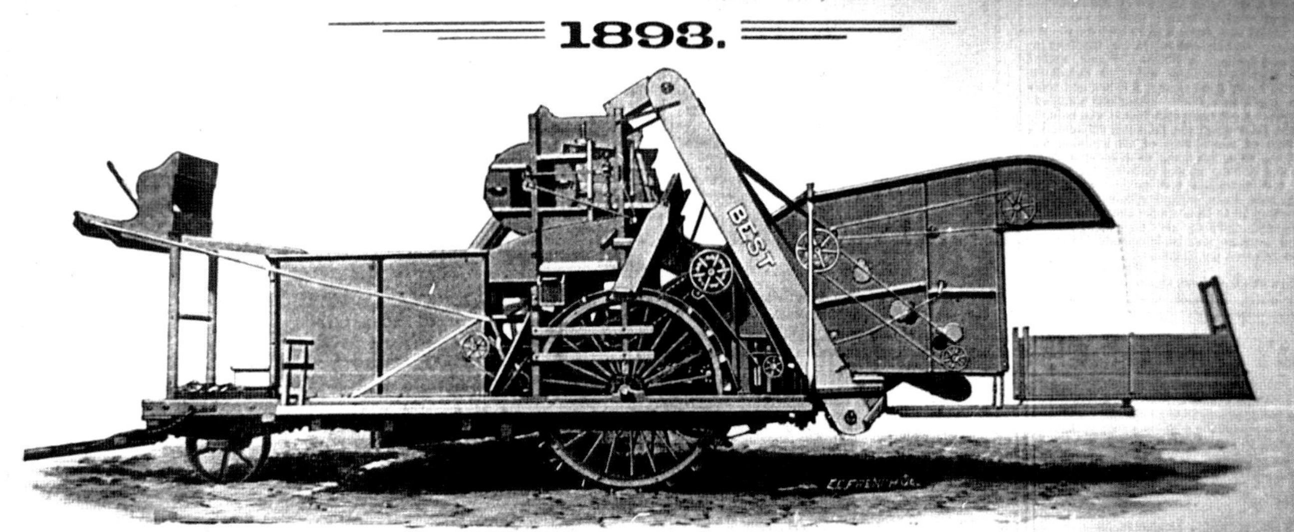

Undisputed Success of this King of the Field in the Harvest of 1891-1892.

It is indisputably the simplest, lightest draft, best made, best grain-saving machine ever introduced. All farmers who have used it recommend it as superior to any Harvester yet in the field. It is single geared, drives from both wheels, has brakes on both wheels, automatic governor on the wind of the shoe, weighs one to three tons less than any other Harvester, no bothersome chaff carriers, no clumsy chains and expensive gearing to break and wear.

The separator and header wheels are so arranged that it can turn a square corner, cutting out a clean corner. It has the best recleaner, has the only header that can handle tangled and down grain successfully, the best separator for handling weedy and heavy grain, has a ten-inch driving belt, new and improved belt tightener, high drive wheels made on the most improved plan,

EVERY MACHINE SOLD GIVES ENTIRE SATISFACTION.

— MANUFACTURED BY —

THE DANIEL BEST AGRICULTURAL WORKS,

Send for Circulars and Prices.

SAN LEANDRO, CAL

Daniel Best's combined harvester. Pacific Rural Press advertisement. 1893. San Leandro Public Library Historical Photograph Collection #01596.

Best horse-drawn sidehill harvester. Circa 1905. San Leandro Public Library Historical Photograph Collection #01571.

From Horses' Power to Horsepower

In the 1880s, western farmers produced a significant tonnage of grain, much of it intended for the export market. Wheat became the grain of choice as it was well suited to the area's climate and soil type. Labor was in short supply and was costly. The progressive western farmers saw increased mechanization as a way to increase production and decrease costs. Farmers' demands for improvements received a ready response from equipment manufacturers.

Harvesters, even the efficient Best Harvesters, still required many horses or mules to power the units. With some harvesters using as many as forty animals, the expenses associated with this type of power were high. Not only were the animals costly to maintain (365 days of food and water and the man power and acreage to provide these needs), but also large hitches were difficult to control and harvest time heat took a considerable toll on the animals.

Power produced by wood and water, not animal muscle and sweat, was being used in the eastern regions of the United States. These steam engines were used to power stationary threshing machines and lumber mills. While some of these engines were traction engines, that is, an engine that could move under its own power, they did not have the capability for pulling a plow or a combine. Pulling capacity (drawbar load) would make its debut in traction engines designed and built in the western United States.

One of the first practical traction engines was built by Marquis de Lafayette Remington of Woodburn, Oregon. As Remington was quoted in the August 16, 1890, *Pacific Rural Press*:

Daniel Best 40 Horse Power traction engine. The wheels 7 feet 6 inches in diameter and 26 inches broad. This engine weighs 9 tons. Daniel Best is shown at the controls. 1891 catalog. San Leandro Public Library Historical Photograph Collection #01588.

Best traction engine hauling gravel at San Leandro. Originally designed for agriculture, Best engines were easily suited for other uses. 1891 catalog. San Leandro Public Library Historical Photograph Collection #01491.

> *In 1885, I turned my attention to the construction of a traction engine, and with the crude appliances at hand in my country shop. I turned out a small, roughly made engine, but one that worked beyond even my own sanguine expectations when put into the woods logging and into the field plowing and to general work. Subsequently, I made another and more complete one. This one, which I named "Rough and Ready" I took to California, arriving at San Leandro with it in March 1888.*[1]

The land on which Remington demonstrated his Rough and Ready engine was in San Leandro near the Best Agricultural Works. Daniel Best saw Remington's engine as a starting point and bought the rights to manufacture this engine on the Pacific Coast. Remington continued to manufacture engines in Woodburn until fire destroyed his manufacturing plant.[2]

1. "The Remington-Daniel Best Traction Engine," *Pacific Rural Press* (August 16, 1890): 9.
2. F. Hal Higgins Collection, Special Collections, University of California Library, Davis.

Ever the innovator, Best, realizing that the future lay with mechanization, saw ways to improve the operation and profitability of the steam-traction engine. While agricultural interests were the driving forces behind Best's refinement of the steam-traction engine, logging and freighting also profited from his improvements. By September 1890, Daniel had been granted sixteen patents with seven related to improving his steam-traction design. The patent granted to him in September 1889 provided for an auxiliary steam engine to be located on the combined harvester to power it. Steam was provided by means of a hose attached to the Best traction engine pulling the combined rig. Previously, power for threshing came from a ground wheel: The threshing speed was limited to the ground speed of the horse hitch. With steam power employed, the size, strength, and freshness of the multihorse hitch no longer mattered.

To create steam for powering an engine, there are two basic components: water and a fuel source. The early Best steam-traction engines used straw because it was plentiful in the wheat fields. Building on this, Best patented a straw-burning attachment that allowed the threshed stalks to be used as fuel for the traction engine that pulled the combined harvester. A conveyor was used to bring the straw from the rear of the harvester forward to the engine to be used as needed by the fireman to maintain a

Best 110 H.P. traction engine under steam. All Best engines employed an upright boiler design. This configuration allowed for maximum flexibility on both ascending and descending grades resulting in free steaming in any position. The upright design placed the firebox under the water-filled boiler maintaining a safe water level over the crown sheet at all times. Upright boilers were also used in cranes, hoists, and fire pumpers. Return flue-style boilers worked well on level surfaces but encountered difficulties in producing steam on grades. Direct-flue locomotive-style boilers worked well on level and uphill operations but due to crown sheet placement, many times needed to back down grades to maintain a safe water level. The upright boiler of the Daniel Best steam traction engine allowed for maximum versatility in nearly limitless applications. San Leandro Public Library Historical Photograph Collection #01609.

Best engine and ore wagons on a railroad loading dock used to load magnesite ore. Daniel Best is standing with his hand on the wheel of first wagon. Note steepness of the dock ramps. November 1905. San Leandro Public Library Historical Photograph Collection # 01607.

hot fire for steam production. This was done while the entire rig continued to move across the field. Wood and coal were good fuel choices, but not readily available. When fuel oil, another efficient source of heat, was easier to obtain, Best modified his engines to allow for oil burning. This type of fuel was used by many logging operations. The Best steam-traction engines were modified to burn oil before the railroads saw the advantages of this type of fuel in their locomotives.

Water for the boiler had to be hauled across the fields in a horse-drawn water tank to the traction engine. Logging and freighting concerns took advantage of a water supply wherever it was found. Whatever the source of the water, there needed to be a way to introduce the water into the boiler. Early forms of water resupply were a steam pump, an inspirator, and an early type of injector. Best saw a need for a more reliable and variable injector and on September 23, 1890, was granted a patent for improvements that allowed the injector to be used at any time with adequate steam pressure.

Even as Daniel Best was working on his patented improvements, he was equipping his shop with the manufacturing equipment needed to produce steam-traction engines. By February 1889, Daniel sold his first steam-traction engine, priced at $4,500, to J. S. Butler of Tehama County in the Sacramento Valley of California. Best delivered this engine and stayed in Tehama County for seven days while he gave instructions on the care and operation of the engine. Butler, already the owner of a Best combined harvester, later wrote to Daniel Best that the "tractor is a grand success, plows 30 acres a day at the expense of $12.50." This engine was used on Butler's 1,000-acre ranch and also for custom plowing. By August 1889, the Best Works could produce one engine per week.

Agricultural needs fueled rapid experimentation and mechanization on the Pacific coast. Other industries were quick to incorporate these designs for their own uses. Logging, freighting, mining, and road building concerns realized the durability and profitability of using a Daniel Best steam-traction engine. The 1891 Best Agricultural Works Catalog noted three sizes of steam-traction engines offered to the public: 30, 40, and 50 horsepower. Best saw the opportunities of offering complete outfits to freighting and logging concerns. One notable company was the Siskiyou Lumber Company. As stated by E. E. Wickersham, these Siskiyou engines were rated at 50 horsepower with 150 pounds of steam, but steam pressure of 200 pounds was not uncommon as it gave these engines a higher output that logging required. Other improvements, in addition to higher steam pressure, were:

> *a superheater inbuilt with the boiler which added to the economy of operation. In addition to the superheater feature, a condenser and water heater was used, first to save water on long runs; second to effect a further economy. These Siskiyou engines displaced 150 horses, mules and oxen. The cost of operation per 10 hour day with load of 40-50 tons freight was $18.75: Engineer--$4.00; Fireman--$2.50; Lube Oil--.50; 1.5 tons coal--$9.75; Depreciation--$2.00.*[3]

Testimonials were popularly used in news articles, catalogs, and advertising. From a January 16, 1890, letter penned by John Roupe, who had been both a fireman and an engineer on straw-burning field engines, to Mr. Daniel Best:

> *As a straw-burner, this engine gives great satisfaction. . . . I got up to 25 pounds of steam from cold water in 20 minutes, and 130 pounds in 30 minutes, and had no difficulty in holding it at the highest pressure when the engine was working up to her full capacity, and pumping cold water into the boiler at the same time. I unhesitatingly say that the Daniel Best traction engine excels any and all other straw-burning engines I have ever seen or run . . . . I take pleasure in recommending your traction engine as a successful straw-burner, as well as wood or coal.*[4]

3. E. E. Wickersham, unpublished manuscript (1940), Special Collections, University of California Library, Davis.

4. F. Hal Higgins Collection, Special Collections, University of California Library, Davis.

Three Best steam traction engines with loaded ores wagons. Copperopolis, California. Circa 1904. San Leandro Public Library Historical Photograph Collection #01488.

As it had happened a number of times in the past, Daniel Best split his inventive genius between multiple engineering and retailing problems. Throughout the 1890s, grain cleaners continued to be refined, as did combined harvesters and steam-traction engines. The sales and service networks were expanded, as were production facilities. Evidence of this growth was shown by the following sales figures: 1887—$42,368; 1889—$108,408; and 1892—$180,755.[5]

5. E. E. Wickersham, unpublished manuscript (1940), Special Collections, University of California Library, Davis.

On the Pacific coast in the late 1800s, horse power continued to be used on many smaller farms and steam reigned supreme for those large operators who could afford it. Although it was a huge step forward in labor savings over the multihorse hitch, steam-traction power still required a minimum of four people: an operator (engineer), a fireman and water wagon, and fuel wagon teamsters.

Ever looking to the future, Daniel Best saw that the cumbersome steam-traction engines were not to be the final step in the evolution of farming and logging power. Others were experimenting with the internal combustion gas engine. These proved to be sluggish performers. The availability of quality fuel was also an issue. Refined gasoline was expensive and in limited supply. Crude (unrefined) oil could be obtained, but wasn't in the form that the combustion engines could use.

On June 10, 1891, Best filed the application for a patent for a single-cylinder, horizontal gas engine. It was a four-stroke (cycle) hit-and-miss style engine. This engine used an electric spark ignition instead of the standard hot tube ignition in common use at the time. Using a gas engine as a power source had many advantages: It was a compact power source when comparing the ratio of horsepower to weight; once started, it required no full-time operator; and it provided a portable power source that could be used for many operations (i.e., irrigation, threshing grains, and machine shop power). A later Best catalog for his "Gas or Gasoline and Crude Oil Engines" stated the advantages very simply: "No Coal bills; No Ashes; No Smoke; No Engineer; No Noise; No Explosion. Simply and Strongly Built. Cheap and Reliable Power. Economical, Safe and Clean."[6]

These engines proved a popular addition to the machinery lines already being produced at the Best Agricultural Works: horsepower harvesters, traction engines, steam-powered harvesters, and grain cleaners. Taken from a December 15, 1891, article in the *Rural Californian*:

> *It is pleasant to record the general air of prosperity and activity that pervades the Daniel Best Agricultural Works. Orders have been pouring in to such an extent that it has been found necessary to run the factory night and day. In order to do the former expeditiously a $2,000 dynamo has been put in to generate the necessary electricity to furnish light.*[7]

According to contemporary accounts from *Notes from the Best Works*, 1892 showed seventy men were employed, many orders for gas engines were being placed from all parts of California, and the plant was running evenings with monthly payroll between $5,000 and $6,000.[8]

Daniel Best's Electric Crude Oil Vapor Engine. Best was an early innovator in the use of electric ignition. This engine design was awarded a first premium at the California State Fair. Pacific Rural Press *September 18, 1897.*
San Leandro Public Library Historical Photograph Collection #01590.

The Best Agricultural Works offered its gas/gasoline engines in a wide range of horsepower from 2 horsepower, to handle household and smaller farm applications, to 40 horsepower, to provide power to operate a sizable manufacturing plant or pumping station.

Testimonials in the product catalog again showed the durability, ease of use, and cost effectiveness of these engines:

> *I am using the engine I purchased from you every day. It develops more power than you claim, for it runs steady and is very economical in its consumption of fuel. I am perfectly satisfied with it. —Frank Olds*

> *Your inquiry in regard to the three-horse gas engine that you sold me is at hand, and in reply will say that the engine*

6. F. Hal Higgins Collection, Special Collections, University of California Library, Davis.

7. Ibid.

8. Ibid.

Best one cylinder gasoline engine powering a bean cleaner. Lompoc Valley, California. 1895. San Leandro Public Library Historical Photograph Collection #01631.

has been in use nearly a year, sawing wood and pumping water. The engine is all you claim for it.... I cheerfully recommend the Daniel Best Engine as being the most reliable one in the market. —J. A. Turnbaugh[9]

Daniel Best continued to patent improvements to his gas engine. An important patent was granted to Best on November 1, 1892. This patent was for a gas engine and generator. This generator allowed the use of crude oil as a viable fuel. The generator was created by placing a large chamber that had been filled with iron shavings or borings around the exhaust pipe of the engine. Heat was transferred from the exhaust pipe to the shavings. When fuel was introduced into the upper end of the chamber, it flowed over the heated shavings, which gasified the crude oil. This vapor was sucked into the intake of the engine and, when mixed with air, was introduced and burned in the cylinder. The remaining oil could be drawn off the bottom of the chamber (a spigot was provided) and be used for lubricating.

The October 15, 1892, *Best Shop Notes* gave an interesting insight into the operations:

The Best Works are preparing for the next season and from the number of orders already received the coming year will be a busy one for the Works and consequently a good one for our town. There was received from the East and placed in position in the new machine shop this week six lathes and one radial drill, which with the present splendid equipment places the Best Works in the front rank of machine shops in the State. In fact there is but one shop of greater capacity, and that is the Union Iron Works of San Francisco. There will be fifty harvesters and about 200 gas engines manufactured during the coming season."[10]

Five Best brothers. Seated left to right: Daniel, Henry, Zachariah. Standing left to right: Benjamin, Nathaniel. 1893. Identified by Frank Best. San Leandro Public Library Historical Photo Collection #01581.

9. Daniel Best Agricultural Works product catalog (1892), 15.

10. "Good News," Best Shop Notes (October 15, 1892): 1; F. Hal Higgins Collection, Special Collections, University of California Library, Davis.

The Best Manufacturing Company

With the Best product line continuing to expand into logging, freighting, mining, and stationary and motive power and with the increasing sales, the Best Agricultural Works name no longer reflected the scope of the business. The new Best Manufacturing Company was incorporated on January 23, 1893. Directors for this new company included Daniel Best, W. S. Peters, C. Q. Rideout, L. C. Morehouse, and J. J. Scrivner. Various shop notes from that time period note that the plant continued to build harvesters and steam-traction engines while supplying the ever-increasing demand for the Best gas/gasoline engines. The business was also credited with providing the first year-round payroll to the City of San Leandro.

On the personal side, Daniel, Meta, and their six surviving children (their oldest child, Frank, died in infancy) continued to live in the house on the corner of Clarke Street and West Estudillo Avenue. The family consisted of five daughters, Lottie Valena, Meta Ethel, Viola Elise, Bessie Ella, Helena Oleta, and one son, Clarence Leo. May 24, 1894, was a dark day for Daniel Best and his family. Meta Best, age forty, died from complications resulting from a surgery. Daniel now had a family to raise alone as well as a business to run.

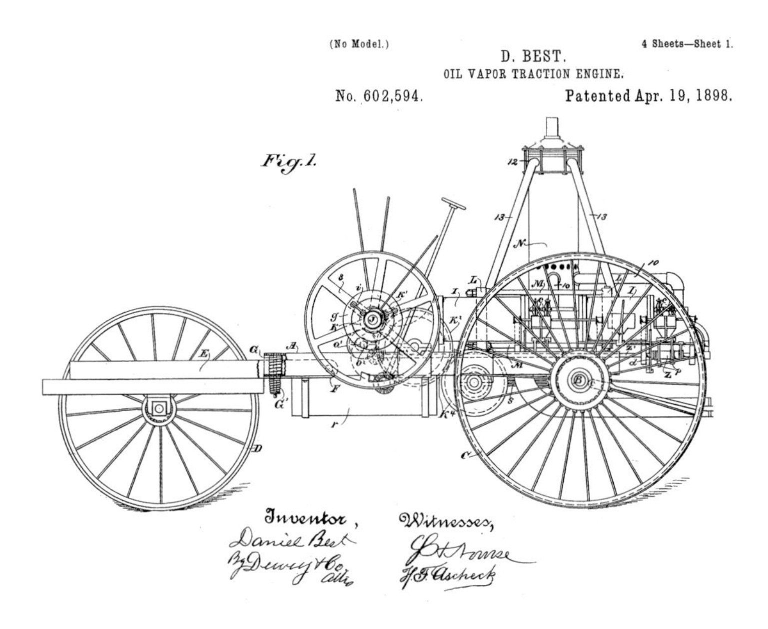

Daniel Best Oil Vapor Traction Engine patent drawing. Patent number 602,594. Patented April 19, 1898.

Daniel Best Oil Vapor Traction Engine. San Leandro, California. 1896. San Leandro Public Library Historical Photograph Collection #01579.

The next years saw the Best Manufacturing Company continue to produce the quality products for which it had become known. Products designed and built in San Leandro found buyers around the world. *The Best Shop Notes*[1] gave some interesting insights as to the work being done at the factory.

"A Profitable Engine" August 18, 1894:

> *Mr. Daniel Best is highly pleased over the result of a test made with one of his gas engines this week. Thirteen gallons of crude oil were used and this amount was sufficient to keep the engine running for ten hours. From this thirteen gallons of crude oil, six and one-half gallons of the lubricating oil was obtained. As the crude oil costs but five cents a gallon and the lubricating oil is worth fifty cents, it demonstrates that a man can make money by running one of Best's gas engines.*

1. F. Hal Higgins Collection, Special Collections, University of California Library, Davis.

Best Shop Notes January 12, 1895:

> *An order has been received to build a hundred horsepower gas engine for a schooner and work will commence on it shortly.*

Best's gas/gasoline engines were now a proven commodity. But the inventor had another use for his engine churning in his creative mind. Given the amount of labor needed to operate and maintain a steam-traction engine and the costs that were incurred, the next logical step for Daniel Best was to incorporate his oil-vapor engine into a running gear that was similar in design to his steam-traction engine. This was reflected with his submission to the U.S. Patent Office of an application for an Oil-Vapor Traction-Engine. The application was filed on September 14, 1896, and granted on April 19, 1898.

According to the *San Leandro Reporter* from July 4, 1896, an interesting event was held at the Best Manufacturing plant:

> *Last Monday afternoon Daniel Best treated his employees to a novel exhibition. The new gasoline traction engine was*

Traction engine tug of war. 1896. Recreation and illustration by Quince Galloway. San Leandro Public Library Historical Photograph Collection #01597.

hitched with large chains to the steam traction engine and both engineers put on full heads of steam. The result was that the gasoline engine hauled the steam engine around the block. This test was another demonstration of the superiority of the gasoline engine over the steam. Some minor changes are being made in the cooling pipes after which the engine will be ready to deliver to the parties for whom it was built.[2]

The following year, 1897, Clarence Leo Best, Daniel Best's only surviving son, officially came into the business as a buyer for the shop. Living within walking distance of the plant, C. L. was certainly no stranger to the operation and layout of the factory. As stated by C. L. himself:

> *From nine years old I traveled free and my Dad hauled me all over the State. I recall this talk with his only salesman, Peter Moy of Livermore about this and that tractor which they visited, mostly if not all, behind some blacksmith shop.*[3]

2. "Tug of War," *San Leandro Reporter* (July 4, 1896): 3.

3. Letter to F. Hal Higgins from C. L. Best (November 5, 1947), F. Hal Higgins Collection, Special Collections, University of California Library, Davis.

Clarence Leo Best. San Leandro, California. Circa 1895. San Leandro Public Library Historical Photograph Collection #01610.

Daniel had this to say about C. L.'s early attraction to the business:

> *Leo grew up in the tractor and farm machinery business. When he was thirteen he showed great interest in the manufacturing end of the business and was constantly at it the entire period that I was active.... Leo took several courses in engineering and also many special courses in his particular line of business. I made him superintendent of my plant at the age of twenty. Before the age of eighteen he was buyer for the plant.*[4]

For further education, C. L. attended the Oakland Business College, the Lick School of Mechanical, and Anderson Academy at Irvington.[5] But by far, the education that served him best was the one found at 800 Davis Street. From the design shop to the foundry, from the machine shop to the assembly floor, and on to sales and shipping, C. L. was exposed to the entire process at an early age. The lesson that products should be built with quality materials, quality workmanship, and pride, then backed by outstanding service, was one that was passed down from father to son. As superintendent for the Best Manufacturing Company, twenty-year-old C. L. was actively involved with the many aspects of the company. He learned valuable lessons from interacting with the farmers, miners, loggers, and freighters who purchased Best equipment. It was the policy for Best machines to be delivered and in top operating order before being turned over to the purchaser. Leo received firsthand the opinions concerning the pluses and minuses of both the Best machines and those of rival companies. The information was brought back to San Leandro, and if needed, changes were implemented. According to a contemporary, E. E. Wickersham, C. L. "never overlooked an opportunity to better himself—prepare for the future. This inherent alertness and originality, plus that rarely found quality, ability to 'sift the good from the bad' made him what he is today."[6]

C. L. and Daniel took a giant technological leap when they entered into the horseless carriage competition. Responding to newspaper accounts of European experiments with the horseless carriage, the Best Manufacturing Company produced a prototype in late 1898. This model was designed to carry eight passengers, furnished with a 7-horsepower, 2-cylinder motor,

4. *Sutter County Historical Society New Bulletin* 3, no. 7 (October 1963): 3.
5. Ibid.
6. E. E. Wickersham, unpublished manuscript (1940), Special Collections, University of California Library, Davis.

and capable of speeds of up to twenty miles per hour. When equipped with water and gasoline for a 150-mile excursion, the vehicle weighed 2,570 pounds. It was reviewed in the March 4, 1899, *Pacific Rural Press*:

> *The success that has attended the manufacture of the road engines built by the Best Manufacturing Co. of San Leandro, Cal., for freighting purposes naturally makes prominent any production of that firm along the line of supplying the traveling public with a device suitable for passenger traffic, and recent trials of a gasoline motor carriage built by that enterprising concern have demonstrated ability to make as great a success of such a motor as has been made in the manufacture of the Daniel Best 50 H.P. traction engine.*[7]

This horseless carriage, while performing well mechanically, had some undesirable side effects:

> *[I]t produced terror wherever it went. The clatter and rattle and motor noise, coupled with the fact that there was "no horse in front" frightened horses, cattle and chickens and paralyzed with fear even women and children. The Bests were particularly concerned over this, and contrived in various ways to overcome it. The final result was the improvising of an exact reproduction of a horse head, life-size, which was attached to the front of the erstwhile horseless carriage. And once more they took to the highway! But to no avail. The horses, cattle, the women and children, were all even more terrorized than before. The Bests, father and son, were despondent.*

[T]he temptation to capitalize this invention and become pioneer automobile builders involved discontinuance of the successful business of making the Best tractor engine. "Leo," said Daniel Best, "suppose you had $4,000.00; would you rather

7. "A California Horseless Carriage," *Pacific Rural Press* (March 4, 1899): 1; F. Hal Higgins Collection, Special Collections, University of California Library, Davis.

The Best horseless carriage with C. L. Best driving. Best Manufacturing Company. 1898. San Leandro Public Library Historical Photograph Collection #01485.

spend it for that infernal machine, which scares horses and children, or take $500.00 and buy a fine span of driving horses?" The son reflected only a few seconds! "I'd rather have the horses!" "So would everyone else," smiled the father, and this disposed of the automobile business as a tempting enterprise.[8]

In the photos showing Leo piloting the automobile, the carriage was filled to capacity with passengers. One of the passengers was identified as Pearl Gray. Pearl Margaret Gray was the adopted daughter of William and Rozzie Gray. W. H. Gray was at one time the secretary-treasurer and bookkeeper for the Daniel Best Agricultural Works. On December 12, 1900, the Best and Gray families were linked when Clarence Leo Best married Pearl Margaret Gray.

Now twenty-two years old, C. L. continued as supervisor for the Best Manufacturing Company and also sold and delivered various products. One story, as related by F. Hal Higgins from an interview with C. L. in the 1940s, showed the creativity of early salesmen:

8. "When Mr. Best Thought Better of It," *Best Tractor News 1*, no. 1 (June 1923): 8.

Best steam traction engine shown hauling 24,000 feet of recently felled logs. J.E. Craney operation, Springdale, Washington. With its upright boiler, a Best engine was favored by loggers. San Leandro Public Library Historical Photograph Collection #01490.

This young fellow had bought the outfit without letting his father know that he was breaking away from the safe and sure oxen and horses. So, when I landed at the mill, way up in the shadow of Mt. Shasta in northern California, the old man called his son into his office and proceeded to give him a bad half hour.... The son came around me and tried to cancel the order. I did a lot of thinking that night without any real sleep as I figured out what to do the next morning. Here I was out to prove to Dad that I could sell his tractors. I was plenty fresh and cocky. But I had a lot of Dad's steel, labor and capital tied up in this deal with 400 miles between the factory and me. Right after breakfast, I fired up the bright new Best 110, hitched on the special Best logging trucks and started driving in circles from mill pond to the woods and back. No one was loading and unloading, just going through the motions of filling my part of the contract. That went on for three hours when the buyer came out and OK'd the deal. You can imagine how I was stepping when I got back to the factory and reported to Dad with a signed check.[9]

The old century was coming to a close, and the future looked bright for the Best Manufacturing Company. But looks could be deceiving.

9. "Steam Doings on the West Coast," *The Iron-Men Album* 6, no. 3 (January–February 1952): 7.

Twentieth-Century Triumphs and Trials

The start of the new century found Daniel Best still applying for and being granted patents. Many of these patents were for improvements to his harvesters and cleaners. Horse-powered harvesters were still popular in many areas of the western United States, and the Best Manufacturing Company continued to produce this type of harvester. In 1902, Daniel was granted a patent for a "side-hill harvesting machine." Other patents issued during the early 1900s related to boiler issues on the steam-traction engines.

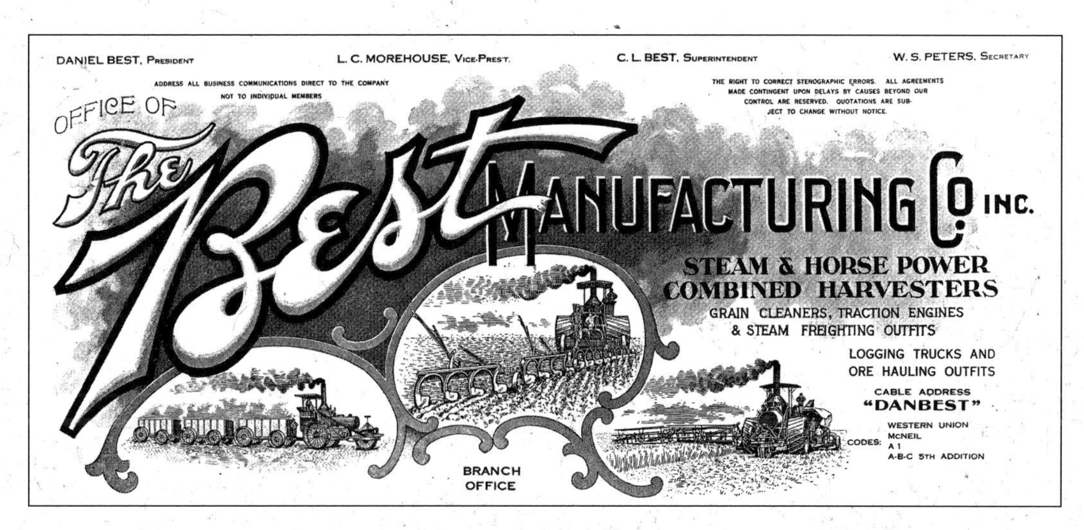

Best Manufacturing Company letterhead. Courtesy of Dan G. Best II.

The excellence of the Best machines was gaining international notice. In a news article from 1903, it was noted that the traction engines were in use in South Africa, South America, Russia, Siberia, and many parts of the United States. The article went on to discuss the anticipated production for the coming year:

> *During the coming season, the company will build twenty traction engines. . . . Sixty horse-power harvesters and ten steam harvesters will be turned out this year. . . . Ten portable cleaners and about one hundred gas engines will also be manufactured.*[1]

Daniel Best continued inventing, manufacturing, and improving his designs, but another Pacific coast manufacturer was traveling along similar lines. Benjamin Holt and his Holt Manufacturing Company were also engaged in the production of harvesters, steam-traction engines, and related equipment. The competition between the two companies proved intense. In May 1904, the rivalry peaked when Daniel Best of the Best Manufacturing Company filed a claim in the U.S. Circuit Court for the Northern District of California (case no. 13,583) asking for a judgment of $100,000 against the Holt Manufacturing Company for patent infringement. The patent in question was number 410307, concerning a secondary engine mounted on a combined harvester to provide independent power to the harvester. The patent application was filed November 22, 1888, with patent rights granted on September 3, 1889.

After three years of pretrial maneuverings, the case started trial on November 21, 1907, with the sixty-nine-year-old Daniel Best taking the witness stand. Daniel explained in detail:

> *I had been working with combined harvesters and improving them—trying to improve them—from 1880 to 1885. In 1886 I went into the manufacture of them. I was manufacturing a horse combined harvester for about three years before I conceived the idea of making them steam combined.*[2]

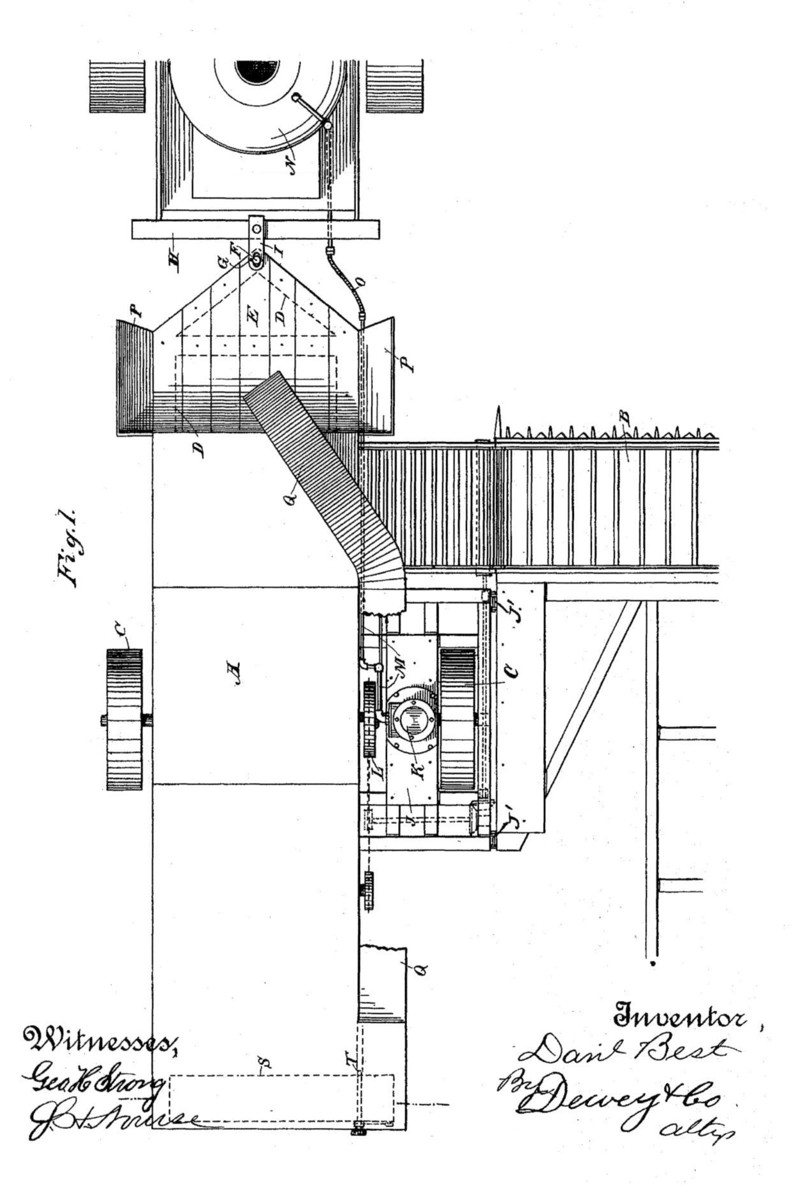

Daniel Best Combined Harvester and Thrasher [sic] patent drawing. Patent number 410,307. Patented September 3, 1889.

1. "Best Works Very Busy," *San Leandro Reporter* (January 24, 1903): 1.
2. *The Best Manufacturing Co. v. The Holt Manufacturing Company*, trial transcripts, case 13,583.

A Best horse-powered combined harvester shown harvesting wheat on the Baird Ranch at Knights Landing, California, in 1902. The Bairds are maternal ancestors of Dan G. Best II. Photo courtesy of Elizabeth Tussey.

When asked by his attorney, J. J. Scrivner, Best gave a detailed description of the operation of a horsepower machine. He also spoke at length on how the reclaimed lands near the Sacramento and San Joaquin Rivers were a perfect fit for his steam-powered combined harvesters:

> *It is the richest land in California and produces the biggest crops, as well as enormous crop of weeds with it. When it drys [sic] up enormous cracks come in that land, from a foot to two feet wide; it cracks both ways, and sometimes it is 6 or 8 feet square; it is impossible to go on it with horses, the horses would fall in those cracks and break their legs; you could not get on them at all... and I told them [the farmers] "Yes, I have a steam combined harvester, we will go in there and cut that grain and save every bit of it."*[3]

Under questioning by his attorney, Best recalled 1888 and his first attempt at building a steam-combined harvester, nearly one year before he filed for his patent:

> *Mr. Scrivner: On your first machine you located the supplemental engine on the front side of the harvester in the same way the defendant [Holt Manufacturing Company] does now?*
>
> *Mr. Best: Yes, in about the same place the defendant does now.*
>
> *Mr. Scrivner: And why did you move it back?*
>
> *Mr. Best: I moved it back because it was more convenient to belt from it and to connect from it on the side of the machine than it was on the front.*[4]

Daniel Best, at the request of Mr. Scrivner, used a model of the patented steam-combined harvester to explain the various parts and operations. Best also explained the differences between a horse-powered combined harvester and his steam-powered combined harvester. The horse-powered rig got its power from a main driving wheel, often called a bull wheel. The ground speed of the horses set the speed of the bull wheel and the speed at which the harvester cut, threshed, and separated the grain. When the grain was heavy, the horses had to pull harder causing them to slow down, which also slowed the harvesting process, often choking the harvester. When grain was light, the horses pulled more easily and faster, which caused the harvester to speed up and exhaust a large quantity of marketable grain with the straw and chaff. If the horse-powered rig stopped outright, all power for the threshing operation ceased. With his idea of a steam-powered harvester, Best was able to supply the harvester with a reliable and constant source of power. His power source received steam from the traction engine's boiler and was not dependent on the ground speed of traction engine. The separate auxiliary engine on the harvester provided the same threshing power whether the traction was moving slowly or quickly or standing still. This consistent power enabled farmers to salvage downed grain at cost savings estimated by Best to be between $6 and $10 per acre.

3. Ibid.

4. Ibid.

When questioned about comparisons between the Best and Holt steam harvesters, Best admitted that as far as the separator, header, and engine, the Best and the Holt machines operated seemingly the same. The contention lay with positioning of the auxiliary steam engine. The Best Steam Harvester had the auxiliary engine mounted on the harvester frame. The steam was carried from the traction engine boiler via a flexible pipe to the auxiliary unit.

But Daniel testified:

> *from 1893 to 1900, Mr. Holt—the Holt Manufacturing Company was building a combined harvester, with their auxiliary engine mounted on the frame of the traction-engine. That is up to 1900. Then they made a change and put their engine on the front side of the harvester.*[5]

Under continued questioning, Daniel affirmed that he had no objections prior to 1900 because Holt

> *did not use any construction of mine. He was perfectly at liberty to build all that kind of machinery, and I claim nothing.*[6]

At Scrivner's request, Best explained how the pre-1900 Holt system carried the power back from the supplemental engine (mounted on the traction engine frame) to the mechanism on the harvester. This was accomplished by using knuckle-couplings and tumbling shafts and gearing. Problems arose when turning corners and going over uneven ground. The shafting would "get out of line, get out of shape and break."[7] After 1900, Holt shifted the auxiliary steam engine off the traction engine frame to the harvester frame and carried the steam back through flexible couplings. Scrivner questioned Best about the placement of the auxiliary engine and its success rate:

> *Mr. Scrivner: Then, as you understand it, after a test of several years, and the making and selling and operating of these machines with the auxiliary engine on the traction-engine frame, it proved to be a failure.*
>
> *Mr. Best: It proved to be a failure, yes sir.*
>
> *Mr. Scrivner: No one to your knowledge had ever done that [harvester-mounted auxiliary engine] prior to your invention, or discovery of this?*
>
> *Mr. Best: No.*[8]

On cross-examination, Mr. I. M. Kalloch, attorney for the Holt Manufacturing Company, had Best clarify many aspects of both the traction engine and the steam-traction harvester. The Berry patent for steam harvester was brought to the court's attention by counsel for the defendant. Best stated that the outfit was not economically successful and that fewer than five were built. He also stated that he had purchased the Berry patent for $1,000.

During the course of the trial, many other noted agricultural inventors were questioned about the design and operation of their respective machines. This list included C. L. Best, Pliny E. Holt, and George S. Berry.

J. H. Davis, called as a witness for the plaintiff, stated that he was currently employed by the Best Manufacturing Company but prior to 1902 he had worked for the Holt Manufacturing Company as a mechanic on and off for thirteen years. Again, the testimony centered on the value of a steam-harvesting outfit and how the placement position of the auxiliary steam engine affected the success of such rigs. Under questioning by Scrivner, Davis testified that in the spring of 1900, he bought a Holt harvester and had a discussion with Ben Holt concerning the auxiliary steam engine:

5. Ibid.

6. Ibid.

7. Ibid.

8. Ibid.

Advertisement for Daniel Best steam traction engine and harvester. 1889. San Leandro Public Library Historical Photograph Collection #01491.

I was figuring on buying the machine, and I suggested putting the auxiliary on the separator. Mr. Holt did not agree to put it on there for some time.... He said that Mr. Best was using this machine that way and Mr. Best had a patent on it and he did not want to get in any trouble. We finally settled the matter, and he says, "we will put it on all right."... I didn't want to get in any trouble over it, so went to Mr. Holt and asked him on the quiet—privately—in regard that the question might come up in regard to a royalty on this machine.... He says, "Damn it, we will put it on and if we have to pay for it, we will pay for it." That is all there was to it.[9]

When cross-examined by Mr. Kalloch, Davis admitted that he quit the Holt Company because he felt that he

9. Ibid.

Best Manufacturing Company. In the foreground is a newly constructed Best Combined Harvester ready to ship. 1904. San Leandro Public Library Historical Photograph Collection #01534.

> *was not getting enough money for my work. They did not discharge me. They did not notify me to quit, I just simply resigned, quit.... I did not feel bitter toward them.*
>
> *Mr. Kalloch: And are you trying to get even with them?*
>
> *Mr. Davis: No sir, they don't owe me anything. I feel friendly toward them now.*[10]

The defense called Pliny E. Holt, nephew of Benjamin Holt, who stated that for a little over eleven years he had been connected with the Holt Manufacturing Company and that he currently was vice president and superintendent. Holt was questioned on whether the Holt factory at Stockton modified the location of the auxiliary steam engine on the harvesters and if it provided materials for farmers who chose to do the modifications themselves. As to the factory changing the location, Holt testified:

> *We have no record of that, no written record, for the reason that all those changes were made on verbal orders; the men who made them were told to go to the storerooms and make their selections of material, and that did not appear on the books in any way, because it was shop work.*[11]

When pressed by Mr. Scrivner concerning the materials used to change the position of the auxiliary steam engine, Holt insisted that all of the materials were stock materials that were carried for wholesale dealers as well as for manufacturing purposes. When cross-examined further, he did admit that some of

10. Ibid.

11. Ibid.

the pieces would only be used when changing the placement of the auxiliary steam engine to the harvester and that some farmers specifically ordered the entire package "to change my upright engine from the traction engine frame back to the harvester."[12]

The verdict was received from the jury of December 19, 1907. After two weeks' time and much technical testimony, the plaintiff, the Best Manufacturing Company, was granted a judgment of $35,000 against the Holt Manufacturing Company for damages relating to copyright infringement.[13]

This was not to be the end of litigation, as an appeal was filed by Holt's attorney on February 15, 1908. I. M. Kalloch charged that numerous mistakes had been made during the trial, and that the Daniel Best patent from 1889 was not a "pioneer" patent. The key evidence cited for the defendant's case was the 1887 patent held by George S. Berry of Tulare County, California, for a steam-powered, self-propelled combined harvester. The Holt attorney contended that the Best harvester contained the same features as the earlier Berry harvester and that the location of the auxiliary engine was only a minor improvement on the existing patent. Kalloch further contended that if Best could make an improvement without penalty, the same right should extend to the Holt Manufacturing Company.

The document filing was completed in May 1908 with both litigants awaiting the decision of the U.S. Circuit Court of Appeals. It wasn't until August 1909 that Judge Gilbert Ross reversed the verdict of the *Best v. Holt* case and remanded it to a lower court for a new trial. While the court system moved slowly, momentous events concerning the two litigants were unfolding.

The patent suit against Holt Manufacturing consumed both time and money, but the Best Manufacturing Company continued with production of its popular equipment. The market was still growing for the horse-powered and steam-powered harvesters, and loggers made the Best steam-traction engine the engine of choice in the Pacific Northwest forests. Best products were also well received in foreign markets. The year 1905 showed sales of $300,000.[14] As noted in the November 4, 1905, issue of the *Oakland Tribune*:

Best steam traction engine and harvester at work in the Knights Landing area of California in 1903. Photo courtesy of Elizabeth Tussey.

> *The Best Manufacturing Co. which now has 135,700 square feet of floor space under roof is preparing to put up another building on the west side of Davis Street. The new building will be 250 × 150 feet on the ground. It will cost about $8,000, and the new machinery which is designed to go into it will cost about $4,000 more. This addition will make in all over six acres under roof. About twenty more men than usual will be used this winter, or 160 in all.*[15]

A steam-traction engine, alone or when combined with equipment for freighting, plowing, or harvesting, was a significant investment for western operators. The 1907 Best Manufacturing Company price list showed Plowing Outfit No. 3 priced at $5,900. This included a Best 110-horsepower traction engine

12. Ibid.

13. Ibid.

14. Terry G. Galloway and Brent D. Galloway, "Daniel Best Biography," *Engineers and Engines Magazine* 15, no. 1 (May–June 1969): 1.

15. "Real Estate Market in San Leandro," *Oakland Tribune* (November 4, 1905): 9.

with drive wheels eight feet in diameter with 60" faces. Also included in the package was a plow cart or tender, a water wagon with the tank capacity of 600 gallons, the oil burning attachment and oil tank, and necessary tools. The steam-harvesting outfits, which included the 110-horsepower traction engine and steam harvester, were priced from $7,900 to $8,300.[16]

After receiving an unfavorable verdict in the patent infringement, the Holt Manufacturing Company was considering possible options to deal with the situation. In a letter dated December 24, 1907, from Holt Company Vice President Ben C. Holt to Pliny Holt, Ben commented on the defeat and how he felt that company attorney I. M. Kalloch allowed information to be introduced that portrayed the company "as though we were more or less guilty." Not wanting an expensive, prolonged appeal, Ben C. suggested that a "favorable deal" could now be made with Best, "owing to the depression which Best is feeling a great deal more than we are." He suggested that if Best was offered a small cash payment, with the balance in long-term, low-rate notes "we could handle this now" and that the Best interests should be contacted "in a quiet way." If this purchase were completed, Holt Manufacturing Company "would be so strong that we could crush outside competition that will keep coming to the surface... this way we would be [prepared] for it because a three cornered fight is a dangerous one indeed, one gets hit from behind too many times."[17]

Daniel Best celebrated his seventieth birthday in 1908. He had invested over forty years of his life to the design and manufacture of labor-saving machinery. The steam power that figured so prominently in the Best Manufacturing Company was losing its position as power of choice to the internal combustion engine. Competitors were experimenting with self-laying tracks that could provide traction without the immense-sized wheels found on steam-traction engines. The costs of maintaining and expanding design, production, and sales staffs and facilities were high for a small-volume manufacturer. The economic downturn of 1907–1908 was affecting Best Manufacturing more than some of the other manufacturers, and years of litigation were also taking a toll. On October 8, 1908, a "group of men in Stockton, all of whom were connected in the Holt Manufacturing Company"[18] purchased the Best Manufacturing Company for an estimated $325,000.[19] Two-thirds interest was to go to the Holt Manufacturing group with C. L. Best retaining one-third interest. All parties agreed to drop the lawsuit that was still in the court system, with both parties assuming their own legal and other costs. The appraisal for the sale contained a detailed listing including $37,000 for the land, various shops, tools, lumber, iron, and completed machinery ($45,100 for traction engines, $42,975 for harvesters) and $30,000 for unexpired patents and machine models, designs, and drawings.[20] The buyout also included an extremely valuable component of the Best Manufacturing Company: Clarence Leo Best.

The *San Leandro Reporter* dated October 17, 1908, reported that Daniel Best had retired:

> *and his son C. L. Best has been elected president of the new organization. The vice-president of the new company is F.O. Meyers and W.S. Peters has again been selected Secretary. The above constitutes the entire change that has been made. The business will be conducted the same as in the past, with the exception that the present heads of the concern will endeavor to make it a larger and better establishment than ever, will put new life and vigor into the business.*[21]

Part of the new life was the introduction of a new harvester and traction engine. The Betty Best combined harvester, which had its patent pending, went into production and was well received. This combined harvester was named for C. L. and Pearl's daughter, Elizabeth (Betty). A round wheel gas-traction engine also went into production. As the Best name was well

16. 1907 Best Manufacturing Company Price List, F. Hal Higgins Collection, Special Collections, University of California Library, Davis.

17. Pliny Holt papers, The Haggin Museum, Stockton, California.

18. "Stockton Man Replaces L. Best," *Oakland Tribune* (February 14, 1910): 1.

19. Reynold M. Wik, *Benjamin Holt & Caterpillar Tracks & Combines* (St. Joseph, MI: American Society of Agricultural Engineers, 1984): 108.

20. Pliny Holt papers, The Haggin Museum, Stockton, California.

21. "Daniel Best Retires," *San Leandro Reporter* (October 17, 1908): 1.

Erecting floor, Best Manufacturing Company, San Leandro, California. Circa 1904. San Leandro Public Library Historical Photograph Collection #01523.

The Best Manufacturing staff poses for a photo. Daniel Best is seated in middle. C. L. Best is in the first standing row, second from left, holding a pipe. Circa 1900. San Leandro Public Library Historical Photograph Collection #00508.

known in freighting circles as producing a well-built, dependable steam-traction road engine, this confidence was transferred to the gas-traction engine.

C. L. carried out duties of the position of president with enthusiasm. The highly skilled workforce that had served Daniel well continued to perform under C. L.'s leadership. Despite the advantages of being a Best from San Leandro, C. L. had paid his dues to the company and had a thorough understanding of what the customers demanded by way of products, what was involved with the production of that equipment, and how extremely important customer satisfaction was to the success of the company. Even though Best tractors and equipment were expensive to build, the purchase price reflected the quality built into the machines. Lower operating and maintenance costs reflected this quality.

C. L.'s skill and knowledge of the operations of the company led to a profit of approximately $90,000 for fiscal year 1909. This accomplishment was praised by Pliny Holt in a letter dated October 5, 1909:

> *I am indeed pleased to note the very good showing that the company [Best Manufacturing] made this past year and there is no reason in the world why they should not continue to do the same and to make from at least $75,000 to $100,000 per year.... There is no use in denying that the round wheel gasoline traction engine is a good saleable machine and will give good results under proper conditions, and until we get the "CATERPILLAR" engine more thoroughly known and introduced, there will be a large sale for that type of machine and the Best company should by all means get their share of the business.*[22]

During 1909, C. L. corresponded with Pliny Holt, who was the president of the Northern Holt Company, located in Minneapolis, Minnesota. He proudly wrote of the new gas traction engines that were in production:

> *California has sure gone crazy over the Gasoline Traction, and it is very seldom that you run across a farmer that talks steam.... The specifications of our new Gas traction are in the hands of the printer, and we will surely mail you some as soon as we receive them.*
>
> *The two Gas Tractions I am building at the present time are using the same frame. The change gear box is practically the selective type sliding gear, operated with one lever, the same as an automobile: three speeds forward and reverse. The slow speed, a mile an hour, the middle speed two and one-quarter, and the high speed, four and one-eighth.*
>
> *We are using the heavy type, slow speed BUFFALO, which I have had re-arranged to suit our conditions; four cylinders, seven by nine and at three hundred and fifty revolutions will develop easily forty-five horsepower. We have listed this engine at four thousand dollars apiece and could sell fifty of them, if we had them.*[23]

While product advancement and production were indeed moving forward, internal political maneuverings were causing dissension in the topmost ranks of the Best Manufacturing Company. In a letter dated January 15, 1910, C. L. Best stated that he

> *very soon discovered that it was the purpose of the majority interest in the company that I should be relegated to the position of a mere figure-head, notwithstanding the large interest in the company held by me, or the important official position I occupied.*[24]

The controversy centered on F. O. Meyers, a Holt employee who had assumed the role of vice president when the Holt Manufacturing Company purchased the Best concern in 1908. Company by-laws were amended to authorize the vice president and secretary, as well as the president, to sign company checks. As stated by C. L. Best:

22. Pliny Holt papers, The Haggin Museum, Stockton, California.

23. Ibid.

24. Ibid.

(Above) *The "Betty Best" horse-drawn combined harvester. Designed by C. L. Best during the Holt-controlled Best Manufacturing era and named for his daughter Betty.* Photo courtesy of Dan G. Best II. (Left) *C. L. Best. 1910.* Photo courtesy of Dan G. Best II.

> *[T]his was supposed to apply to checks which had to be drawn in my absence or under conditions where it was inconvenient for me to perform that duty, but it has been so construed in actual practice, that Mr. Myers has assumed the right to sign all checks of any importance without my knowledge or without consulting me as to the propriety of so doing. He [F. O. Myers] has deceived and misled me so often, that I regret to say that I have no confidence whatever in his personal honesty.*[25]

When the 1910 business year began in September 1909, Meyers was given control of the buying and selling for the company. Concerned with the manufacturing department not receiving materials needed in a timely fashion and thus causing production to fall behind its schedule, Best commented that

> *[Meyers] management has been negligent [and that Meyers] has publicly boasted that he would see that the material did not get around until he saw fit for it to start; so that whoever may be entitled to the credit for the success of last year, I do not propose to take the responsibility of failures for the coming year which I predict.*[26]

In his letter to the Board of Directors of the Best Manufacturing Company, C. L. Best cited other instances of what he felt was an usurping of the powers of his presidency, which resulted in his being little more than a figurehead in the company. The high hopes that Best held upon entering the agreement with the Holt Manufacturing Company failed to be realized. The final paragraph of that January 15, 1910, letter would set into motion tremendous changes for the tractor manufacturing world:

> *I cannot in the form of a letter, recite the innumerable instances occurring daily, of improper conduct, neglect of duties and business, want of efficiency and business comprehension of what should be done, or when or how it should be done, especially on the part of Mr. Meyers. Suffice it to say that owing mainly to his want of ability as above indicated, and to his supercilious and pompous deportment, not only towards me, but to customers and others, I have lost confidence in the whole management. Under such conditions, I cannot see how I can participate in what I regard as such unbusinesslike methods. I therefore herewith tender you my resignation as president and superintendent of your company.*
>
> *Wishing you every possible success, I remain yours respectfully,*
>
> *—C. L. Best*[27]

25. Ibid.

26. Ibid.

27. Ibid.

THE C.L. BEST GAS TRACTION CO.

The world in the first decades of the twentieth century was changing rapidly. Men with vision and determination spurred the technology that set the stage for remarkable advancements. In rural areas, mechanization lessened the amount of man power and time required for farmers to produce and harvest a crop. While these improvements were initially designed with the farmer in mind, other industries were quick to see the far-reaching advantages of mechanical power. Pacific Northwest timber barons saw their profits increase and costs decrease with the expanded use of tractors. Freighting companies were able to provide economical prices for hauling goods, which in turn gave consumers an increased buying power.

The demand for well-built and well-maintained roads was fueled by commercial needs and those of the general population. In California, automobile ownership stood at 780 in 1900. The year 1910 brought that number to 43,210. By 1915, the total automobiles owned by Californians more than tripled to 156,795.[1] West Coast manufacturers rose to the challenge of building the equipment necessary to fulfill those ever-expanding demands for good roads.

After tendering his resignation on January 15, 1910, C. L. Best considered himself free of the Holt-controlled Best Manufacturing Company, but the company had yet to accept that resignation. Best's plans for the future would assure that the management of the Best Manufacturing Company would be forced to accept his resignation.

1. National Transportation Library, U.S. Department of Transportation, Washington, D.C.

BEST WILL START BIG PLANT IN ELMHURST

Acquires Ten Acres Near Railroad Tracks for Proposed Manufactuary
Local and Eastern Capital is Interested
Young Inventor Quits San Leandro Concern to Go Into Business for Self

That headline from the February 18, 1910, *Oakland Tribune*[2] left no doubt as to what C. L. Best saw as his future. Negotiations for the land were complete. The site was located between the Southern Pacific mainline and a branch line running to Stonehurst, providing the means to bring materials in and ship completed products to market. The building permit was issued on March 27, 1910, to C.L. Best Gas Traction Co. for a one-story factory, located at the northeast corner of Quince Street and the Southern Pacific tracks. The value listed on the permit was $8,000.[3] A local newspaper described the production building as

> *[a] mammoth plant for the manufacture of gasoline traction engines and later harvesters. A modern galvanized building, 100 by 300 feet, equipped with the latest in machinery and labor saving devices will be built at once. By summer, it is expected to have at least 60 men employed, and at least 150 in the winter.*[4]

On February 14, the *Oakland Tribune* reported:

> *This morning, George Cowie, formerly superintendent of the Holt Manufacturing Company, of Stockton, replaced Leo Best as general superintendent of the Best Manufacturing Company in San Leandro. A coterie of men who owned a minority of the stock in that [Holt] company became holders of a majority of the stock in the San Leandro concern, with the result that they have become more actively interested and the accession of Mr. Cowie as manager is in furtherance of their plans for a bigger work in the San Leandro plant. Mr. Best is one of the best known men in the state and has a very high reputation among business men in his particular line of work.*[5]

Other area newspapers noted that "Young Best" was reported to have gone East. One speculated that he was purchasing machinery for the new plant, another that he was courting eastern investors for additional capital for the new company. While in the East, he was purportedly looking into the record of patents held by the Holt-owned Best Manufacturing Company and to file new patents of his own. W. S. Peters, secretary of the Best Manufacturing Company, stated in an *Oakland Tribune* article that Best could not "engage in the manufacture of any of the products turned out by the local concern because of the patent rights held by this company [Best Mfg.]." The same article, quoting an "individual who stands high in the council of Best" says:

> *There is no need of infringing on any of the patents held by the Holt people. He [C. L. Best] has plans which will permit of his manufacturing both combined harvesters and gasoline engines which will be entirely clear of any infringement of patent. Young Best practically invented and designed the present type of machine turned out by the San Leandro plant.*[6]

Financing was an issue that had to be resolved in short order. Having resigned from the Best Manufacturing Company, Best forfeited his monetary interest in that company, reported to be $50,000. A major investor in the fledgling C.L. Best Gas Traction Co. was C. L.'s brother-in-law C. Q. Nelson, from Woodland, California, who supplied $25,000 of capital, along with B. G. Peart and other farmers from the Woodland

2. "Best Will Start Big Plant In Elmhurst," *Oakland Tribune* (February 18, 1910): 13.
3. "Building Permits," *Oakland Tribune* (March 27, 1910): 40.
4. "Best Will Start Big Plant In Elmhurst," *Oakland Tribune* (February 18, 1910): 13.
5. "Stockton Man Replaces L. Best," *Oakland Tribune* (February 14, 1910): 1.
6. Ibid.

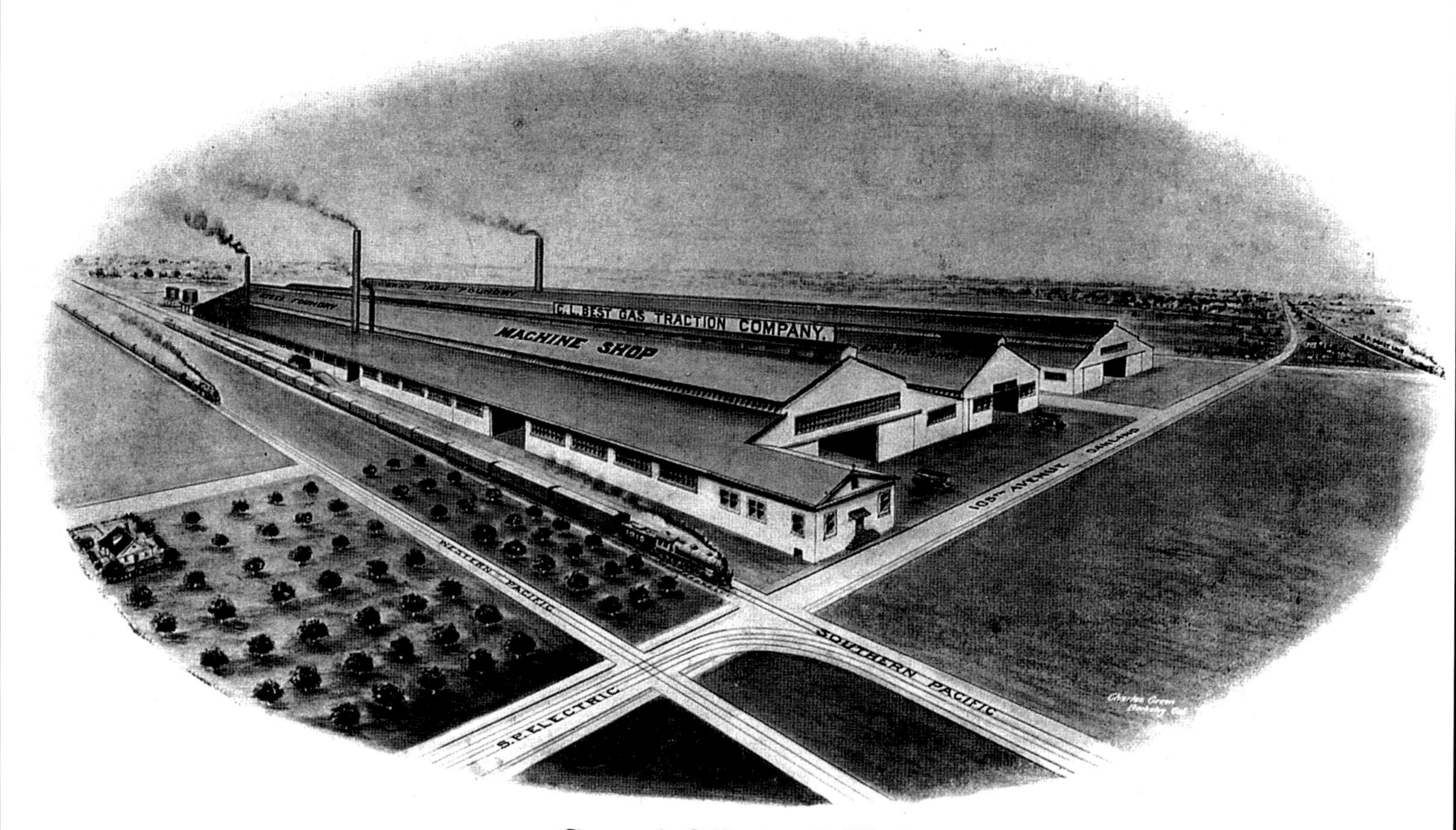

General Office and Works
C. L. Best Gas Traction Co., Inc.
Elmhurst–Oakland
California
P. O. Station G
Phone Elmhurst 130

C.L. Best Gas Traction Co. factory in Elmhurst, California. Company catalog "E" 1913. Author's collection.

area contributing another $25,000. The final third of the setup amount was reported to have come from Daniel Best, although he was not listed in any financial documents as an investor. The elder Best had done a robust business with farmers in the Woodland area. His grain cleaners, harvesters, gas engines, and steam traction engines found ready acceptance with the progressive and successful farmers of that area. Expectations for C. L.'s new venture were high.

The news report that the articles of incorporation of the C.L. Best Gas Traction Co. had been filed with the county clerk also stated that the purpose of the new company was to "manufacture, deal in and sell all kinds of harvesting machinery, traction engines, automobiles, etc."[7] The board of directors for the first year was listed as C. L. Best, Chas. Q. Nelson, B. G. Peart, E. B. Stone, and J. J. Scrivner. According to the *Woodland Daily Democrat*, both Nelson and Peart were frequently listed as taking the train south to check on their investments.

While it could have been easy to be consumed with the day-to-day problems of starting and running such an ambitious venture as the C.L. Best Gas Traction Co., C. L. Best had a rich background from which to draw strength and knowledge. The boy who followed his inventor father into the wheat fields of the Pacific Northwest saw firsthand the hardships that farmers faced bringing a profitable crop to market. His analytical mind followed the processes used by Daniel to determine the surest way to resolve those problems: innovative engineering, quality materials, reliable customer service, and a bit of trial and error.

Being the son of Daniel Best assured access to the Best Manufacturing shops, and C. L. learned from each worker—draftsman, foundry man, assembler, salesman, and all of the others—that every part of the process was important to the finished product.

He had seen the progression from horse power to steam power. The next step was the internal combustion engine. Farming, as well as logging and freighting, demanded refinements of the tractor manufacturers. The future held seemingly endless possibilities.

7. "Incorporation of Best Traction Company," *Oakland Tribune* (March 26, 1910): 12.

But a state-of-the-art manufacturing facility, established financing, and leadership with a vision of the future required a fourth component to complete the leap into a viable business concern: a saleable product. C. L. Best had been instrumental in the success of the gas-traction engines designed and produced at the Best Manufacturing Company in San Leandro. He was well aware that the days of the steam-traction engine were coming to a close and gas traction would be the choice of forward-thinking farmers, loggers, miners, and freighters. Best determined that his new C.L.B. tractors would embrace the future and use the internal combustion engine for power.

The initial offerings of the C.L.B. gas tractors were at twenty-five, forty, and sixty horsepower. These tractors featured a "Buffalo Gasolene [sic] engine."[8] Using an engine built by an established manufacturer allowed the Best Gas Traction Co. to get its tractors on the market more quickly. The 1910 advertisements, as well as sales catalogs, stressed the strengths of the machines:

> *The C.L.B. is mounted on springs, front and rear. The C.L.B. has all steel gearing encased and running in oil, it has three speeds forward and [one in] reverse, its valve mechanism is thoroughly protected from dust. It is equipped with Bosch dual ignition and double Schebler carburetors and mechanically driven oiler that reaches all parts. These are some of the features coupled with its immense power and its economical fuel expense that helps make the C.L.B. . . . THE TRACTOR IDEAL.*[9]

Advertisements in newspapers and agricultural magazines were used to set C. L.'s new company apart from the Best Manufacturing Company in San Leandro. The ads made blatant references to the "other factory":

> *The C.L. Best Gas Traction Co. is located at Elmhurst, Oakland. Do not confound it with any other factory for the*

8. C.L. Best Gas Traction Co. advertisement, *Woodland Daily Democrat* (August 30, 1910): 7.

9. Ibid

Best 70 H.P. Round-Wheel tractor. *The first Best tractors were rated at 70 horsepower and later raised to 75 horsepower.* Company catalog, 1915. Author's collection.

> *C.L.B. factory, like the C.L.B. engine is in a class by itself. Experience has perfected our product, established our standard, made our reputation and proven our guarantee.*[10]

The new C.L.B. tractors were well received, but buyers were cautious of the new products, and many required demonstrations in the field. From the August 30, 1910, *Woodland Daily Democrat*:

> *On Monday there arrived from Elmhurst a gas tractor engine manufactured by the C.L. Best Gas Traction Company. Supervisor Scott has taken the engine out to the power line lane for demonstration. If it works satisfactorily, and the company is willing to furnish a guarantee that it will, Supervisor Scott will probably purchase it for use in road construction work. It is a four-cylinder, forty-five horse power engine and can be used to haul plows, scrapers, barrows, etc. C.Q. Nelson and B.G. Peart of this city are interested in the factory where it is manufactured.*[11]

The prosperous ranchers in the Woodland, California, area were supporters of Daniel Best's horse- and steam-powered

10. Ibid.

11. "Tractor Engine Arrives," *Woodland Daily Democrat* (August 30, 1910): 6.

harvesters and steam-traction engines. Quick to realize the increased profitability and ease of use of gasoline-powered units, ranchers turned to the C.L.B. tractors and the arrival of new machines was reported in the daily paper:

> LATEST TYPE ENGINE
>
> *One of the C.L. Best's brought to Woodland*
>
> *The north-bound freight this morning brought in two of the latest type traction engines from the factory of the C.L. Best Gas Traction Company, of Elmhurst, in Alameda County. One of them, a six-cylinder 60 horsepower machine, will go to Driver & Murray of Knights Landing, where it will be used on the old Fair Ranch, adjoining that part of the property farmed by Schoeller Bros., who recently bought a 40 horsepower C.L. Best. The other, a 4 cylinder 25 horsepower, will go to Ben H. Stephens at his ranch near Madison.*
>
> *The "C.L. Best" is a powerful looking engine and creates more than ordinary interest among our people from the fact that its designer, Leo Best, is well and favorably known here and that a goodly amount of Woodland capital is centered in the enterprise at Elmhurst...from the looks of the machines unloaded today, their success is predicted. They have our best wishes.*[12]

The quality of the products offered by the C.L. Best Gas Traction Co. was directly affected by the quality of the materials used in the manufacturing process. Grey iron (cast iron) was commonly used by tractor manufacturers. Confirmed by his years involved with design, production, and after-sale service, Best was aware of the limitations of cast grey iron and the superiority of cast steel in tractor production. But how could he assure the quality, availability, and affordability of the steel needed for his C.L.B. tractors? The answer was the C.L. Best Steel Casting Co. Best newspaper advertisements touted the fact that:

> *We make steel castings. Our factory is equipped with a Bessemer Process Steel Converter of 4,000 pounds capacity. We are the only manufacturers on the Pacific Coast making our own steel castings.*[13]

The use of steel was featured in many newspaper advertisements for the C.L.B. gas tractors. Calling "cast iron a poor substitute for steel," the ads from early 1911 focused on a long-held Best belief that the purchase price of a tractor wasn't an accurate way to evaluate what the actual cost of the tractor would be over its working lifetime:

> *Cheap machinery like cheap shoes and cheap clothes wears out soon, while the better quality not only lasts longer but gives better satisfaction while in use. Cheap Traction Engines can be built for short lived troublesome service, but the wise engine buyer prefers to pay the additional price required for high class machinery. Ask the man—he will tell you how small his fuel expense is, and he will also tell you that the C.L.B. is the highest grade gas traction engine in the field today.*[14]

One way of showcasing a company's products was demonstrating them at a fair. Held at the state and local levels, these events drew large crowds and were an important means of displaying new machines and generating public interest. The C.L. Best Gas Traction Co. was a participant in many of these fairs. According to the September 2, 1911, *San Leandro Reporter*:

> *The gas traction engine exhibited by the C.L. Best Gas Traction Engine Co. of Elmhurst at the California State Fair in Sacramento was awarded the first grand prize last Thursday afternoon by the prize committee, over all other competitors. There was a very strong competition for the first prize and the Best Engine's victory shows its superiority*

---

12. C.L. Best Gas Traction Co. advertisement, *Woodland Daily Democrat* (December 6, 1910): 6.

13. C.L. Best Gas Traction Co. advertisement, *Woodland Daily Democrat* (March 14, 1911): 7.

14. Ibid.

(Above) *C.L.B. Electric Combined Harvester. Company catalog "E" 1913.* Author's collection. *C.L.B. Round Wheel and Track Laying* (Left) *Tractor logo.* Company catalog "E." Author's collection.

over all other makes. The model exhibited is one of the latest and is rated at sixty horsepower.[15]

The year 1913 again saw the C.L.B. tractors entered in the California State Fair in Sacramento. The company received three awards: first premium & gold medal for the 70 H.P. Round Wheel Engine, first premium & gold medal for the 70 H.P. Track Layer Engine, and second premium and silver medal for the 25 H.P. Round Wheel Engine.

Best offered a new product that was featured in advertisements and at the 1911 California State Fair: the C.L.B. Electric Combined Harvester. Since the C.L. Best Gas Traction Co. built what its name stated, gas-traction engines, C. L. needed to find a power source for the harvester that did not involve steam as was used in the past. His answer was an electric motor. Similar to the steam-combined harvesters in that an auxiliary engine was used to provide power to the harvester, Best powered his auxiliary engine with electricity from a generator powered by a gas tractor. This enabled the ground speed of the tractor-harvester unit to be varied according to field conditions, while the electric motor provided constant threshing power to the harvester. Another plus to this design was that, providing a gasoline-traction engine was used to pull the harvester, existing models of steam- and ground-driven harvesters could be adapted to the new electrical system.

The C.L. Best Gas Traction Co. continued to build the electric harvester for about two years and then sold the design to the Harris Harvester Company. The Best Company focused on round-wheel tractor production, but bigger things were on the horizon for C. L. and his company: things that would test the man, test the company, and test the tractor that was soon to be unveiled to the world.

C. L. BEST, President | CHAS. Q. NELSON, Vice-President and Treasurer | B. G. PEART, Secretary

STANDS FOR QUALITY

C. L. Best Gas Traction Company

Manufacturers of

Gas Traction Engines,
Gas Road Rollers
Electric Combined Harvesters
and Steel Castings

P. O. Box 358 - - - - - - - - - - - Phone Elmhurst 130

Office and Factory at

ELMHURST - - - - - - - - OAKLAND, CALIFORNIA
105th Avenue

C.L. Best Gas Traction Co. catalog. Spring 1911. Author's collection.

15. "Local Engine First In Sacramento," *San Leandro Reporter* (September 2, 1911): 4.

Tracks According to Best

Farmers on the Pacific coast embraced the cost savings and practicality of the tractor: first steam powered, then gasoline powered. As horses and mules were retired, large crews of workers were no longer required to produce a crop. Vast acreages were put under plow. Tractor manufacturers were shaping their products to meet the demands of the farmers.

The large, round-wheeled tractors that were being produced did an adequate job, but had inherent drawbacks. The great weight of the tractors required ever larger wheels to maintain sufficient traction in rainy weather or in soft, marshy, or sandy soils. The broad wheels led to difficulty in maneuvering the machine. The large wheels also made the tractors impractical for orchard work. Local farmers were aware that continued use of these machines produced severely compacted soils that diminished yields.

As early as 1907, C. L. Best became interested in providing a better means of traction. The concept of putting a track around the wheels and using truck rollers to carry the weight of the tractor was being tried. Best soon realized that while the fundamental principle of that type of self-laying track system was sound, improvements to the design would need to be made before it could be marketed to farmers with confidence. During his tenure as president of the Best Manufacturing Company, now owned by the Holt Manufacturing Company, C. L. had high hopes of applying his ideas to the existing and future product lines. How acceptable those ideas were to the new management team soon became apparent: Best resigned from the San Leandro company and the C.L. Best Gas Traction Co. came into existence in February 1910.

Best's ball-tread tractor on a Western Pacific railway flat car. This tractor was C. L. Best's first attempt at a track-laying tractor. Special Collections, University of California Library, Davis.

Even though the round-wheel tractors proved profitable for the newly formed company, Best proceeded with the research and development of a gasoline-powered self-laying track engine. His initial track-laying tractor was completed in the spring of 1911. This tractor, commonly referred to as a "ball-race track-laying tractor," was sold and shipped to Miller & Reveal at West Lodi, California.[1] While the tractor performed as designed, the

> *ball-bearing arrangement proved expensive, grit and dirt keep the tractor from service and inequality of hardness in [an] occasional ball, caused them to wear unevenly with undesirable results.*[2]

It was during his annual two-week vacation and deer hunt (early August 1911) that Best decided upon changing the anti-friction bearing-balls to bearing rollers. Upon his return to the Elmhurst plant, he "laid out said tractor in all its details" and "prosecuted the construction of such a track-layer diligently and in accordance with the usual course of business."[3] C. L. Best's son, Daniel G. Best, remembered an incident involving hunting and tractors:

> *Dad could never stop talking about building tractors. One time, he and I were in a duck blind. He kept talking about something "they" were having trouble with, trying to figure it out. I finally got mad and said, "Are we here to hunt ducks or build tractors?" "Okay, okay," he said. "We're here to hunt ducks." But that didn't last long. Soon he was building tractors again.*[4]

Newspaper advertisements from late 1912 focused on the "New C.L.B. 60 H.P. Gas Tractor," touting it as "perfected, all-steel and backed by a year's guarantee. A home company—a California product."[5] To whet the interest of local producers, the same advertisement promised a demonstration in the Woodland area within forty days:

> *The new C.L.B. 70 H.P. "TRACK" Engine—something better, more surface, more power, more serviceable, less upkeep. Absolutely all steel, all enclosed, mounted on springs. For further information, see our Yolo County representative.*[6]

Crucial to the Best concept of a self-laying track was his development of the hinged radially oscillating truck, also known as an oscillating roller frame. This design, Best felt, was far superior to the "absolute rigid frame carrying the entire mechanism"[7] construction used on other track-laying tractors. By allowing the individual track unit to follow the major contours of the ground, while bridging any minor unevenness, the oscillation permitted the Best machine to maintain its power over these uneven surfaces.

In his advertisements, C. L. Best used comparison to a railway locomotive as the means of explaining the workings of the track-laying tractor. While the railway engine had to travel on

1. C. L. Best deposition (March 20, 1916), U.S. Patent Office documentation.
2. Ibid.
3. Ibid.
4. Personal interview with Daniel G. Best (April 6, 2006).
5. C.L. Best Gas Traction Co. advertisement, *Woodland Daily Democrat* (December 12, 1912): 7.
6. Ibid.
7. *Holt Manufacturing Company vs. C.L. Best Gas Traction Co.*, C. L. Best deposition (August 31, 1915): 4.

C.L.B. 70 H.P. Track Engine No.84. Due to the interspersing of tractors (round-wheel and track-type) in the serial number order, No. 84 was not the 84th 70 H.P. tractor manufactured. In all probability it was the first all Best TrackLayer produced. Any TrackLayer built before No. 84 would have had the 80 horsepower Buffalo engine and production would have been very limited. The 70 H.P. and 75 H.P. tractors used differential steering. This concept would reappear on Caterpillar tractors in the mid 1980s. Company catalog "E." 1913. Author's collection.

immoveable rails, the track-laying tractor could pick up and put down its "rails" or tracks. Other general selling points for self-laying track tractors were not compacting farm soils, stability on side hills, sure footing on forest undergrowth, the ability to traverse muddy and slippery soils, and ease of transition between those slippery soils and dry areas.

Though Best's concept of the tracks and related assemblies was later proven correct, as with any new product, there was a period of refinement. A number of individual parts made up the track unit. The link pins that provided the pivoting points of the tracks were crucial. These pins caused many problems for track-type tractor manufacturers. C. L.'s patented invention, the rocker joint, was billed as "causing minimum friction, needing no greasing and extending the life of the tracks."[8] While the rocker joints did prove serviceable, Best was not satisfied with their performance and continued to work toward a solution. It was not until 1920 that C. L. Best perfected the design that would be the industry standard. But four years earlier, in 1916, the Best Company was ready to make other important changes.

8. C.L. Best Gas Traction Co. advertisement, *Fresno Bee* (September 24, 1914): 13.

Truck wheels, drive sprocket and track extended.

Spring mounted "BEST" Oscillating Track.

The entire track assembly swings on axle "B" of the drive sprocket wheel. The front idler is mounted on pivot "C." Each assembly of track, idler, sprocket wheel and set of truck wheels swing independent of the assembly on the other side of the machine. Spiral springs "A" carry the weight, so that the tractor is spring mounted like a Pullman coach. "D" is the main frame and "E" is the tractor truck frame flexibly hinged to axle "B."

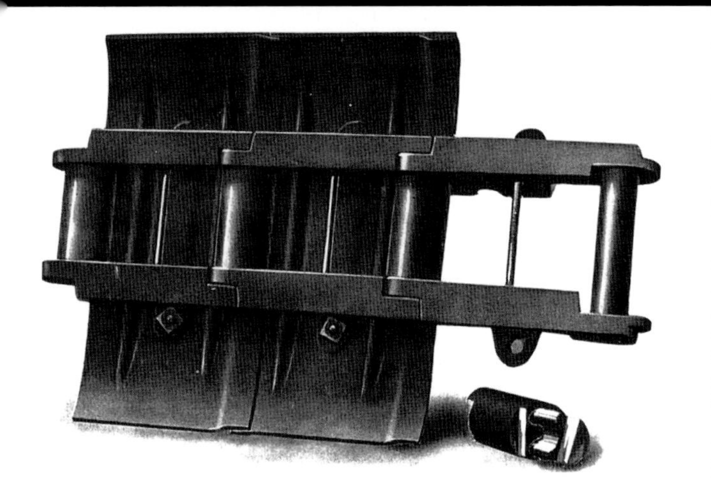

(Lower left and lower right) *Two views of C. L. Best's patented rocker joint shown positioned in the track and separately.* Company sales literature. Author's collection.

Making Tracks to San Leandro

The first track-laying tractor Best offered to the public was the 70 H.P. TrackLayer. Built at the Elmhurst plant, this tractor was well received when shown at the popular agricultural events that showcased new equipment. The California State Fair of September 1913 saw the Best 70 H.P. TrackLayer Engine receiving a first premium and a gold medal. Later in production, the spelling of TrackLayer evolved into Tracklayer, the brand name for the C.L. Best Gas Traction Co.'s track-type tractors.

While production was increasing at the C. L. Best plant in Elmhurst, the City of San Leandro was soon to lose one of its oldest manufacturing firms. The Best Manufacturing Company, originally started in 1885 by Daniel Best, had been purchased by the Holt Manufacturing Company of Stockton in October 1908. For the next five years, the company successfully built various harvesters and round-wheel tractors.

June 28, 1913, saw the local newspaper, *The San Leandro Reporter*, lamenting on the closure of the Best factory. Round-wheel tractor production was to be discontinued, while harvester production was being moved to Stockton:

> *Now they are about to move these shops that have advertised San Leandro all over the world. It is a severe loss, as it has been one of San Leandro's greatest assets for years. Just what will become of*

the property and whether the buildings will be destroyed or allowed to remain is not known at the present time.[1]

The same edition of the paper spoke of the loss of the C.L. Best Gas Traction Co. to Elmhurst:

He [C. L. Best] is meeting with as much success as his father and is employing a large number of men. It is to be regretted that he did not locate his shops here in San Leandro, but lack of encouragement and inducement lost for San Leandro this concern, and Elmhurst is now reaping the harvest that San Leandro should have had.

As the raw materials and dismantled manufacturing machinery were making their way to Stockton, workers were conducting the final closing of the plant. The Holt Manufacturing Company was also tying up any loose ends concerning the Best Company. One final legal issue was addressed:

Deeds were filed at the office of the county recorder yesterday by which all rights, title and interest in the property of the Best Manufacturing Company, makers of farm implements at San Leandro, were transferred to the Holt Manufacturing Company. The filing is believed to be entirely formal, the plant having been sold by the Best people to the Holt concern about five years ago.[2]

When it purchased the Best business, the Holt Company acquired the patents held by that business. When items were submitted to the Patent Office for consideration, a physical model was included along with the line drawings. The models associated with the patents of Daniel Best and the Best Manufacturing Company were housed in a masonry building located on West Estudillo Avenue. As told by Dr. Terry Galloway, great-grandson of Daniel Best:

"BEST"

"TURNS OVER THE EARTH"

PLOW TABLE

Basis of 10 hour day, pulling 10 inch plows, turning 6 inch furrows at a speed of 2½ miles per hour

SOIL	NINETY H.P.		SEVENTY-FIVE H.P.		FORTY H.P.		THIRTY H.P.	
	No. Plows	No. Acres	No. Plows	No. Acres	No. Plows	No. Acres	No. Plows	No. Acres
Light . .	20	45	16	35	10	22	8	18
Medium .	16	35	12	27	8	18	6	12
Heavy . .	12	27	10	22	6	12	4	9

These are conservative figures. The Tracklayer will pull much larger loads but we advocate underloading

Table of recommended work loads for Best Tracklayers. Company catalog. 1917. Author's collection.

I asked my grandmother about the closing of the old plant. It was a sad time in San Leandro. She told me about the patent building, the one housing all the models that her father used when he applied for his tractor and harvester patents. That building was located where the BART station now stands in San Leandro. She said that the last thing the Holts did when they left San Leandro for good was to burn the patent building. Now remember, this was a masonry building—fireproof. The only way to get it to burn up would be to use kerosene to start the fire and help it along. She felt that it wasn't an accidental fire.[3]

Meanwhile, the Elmhurst-based C.L. Best Gas Traction Co. was growing. Tracklayer production at the plant was about one per month. But in late 1913, Oscar Starr, who had been

1. "Best Plant to Close," *San Leandro Reporter* (June 28, 1913): 1.
2. "Holt Company Files Deed to Plant," *Oakland Tribune* (July 13, 1913): 1.
3. Personal interview with Dr. Terry Galloway, great-grandson of Daniel Best (July 11, 2009).

employed by the Holt Company's Aurora Engine Works, left the Holt concern and went to work for Best. Starr was the head of production for the Best plant. Through his expertise and guidance, the one hundred or so men employed in the Best plant increased tractor production to over eight tractors per month.[4]

As with many newly formed businesses, finances were a struggle: Payroll was due monthly; material bills had to be paid; shop, office, and advertising expenses were growing; and money was required for research and development of new products. If sales didn't produce the funds required, a trip to a lending bank was in order. With Oscar Starr heading production, C. L. Best could focus on growing his company.

Many of the Best catalogs stressed the use of high-quality metals in production of the track layers:

> *Where mushy cast-iron might be used and might be satisfactory, close-grained hard iron was selected; where crank bearings might get along with a lead base, a tin base was used to secure better results. And so it is all the way through.*[5]

As the way to control the quality of the metals used, C. L. Best had started his own steel converter and foundry. Outside demand for steel resulted in the forming of a subsidiary company, the Best Steel Casting Company. This action proved profitable, and the foundry size had to be increased a number of times to accommodate the increase in business.

In 1915, with an eye on the agricultural market share controlled by Henry Ford, General Motors had a report on tractor production compiled by independent engineer Philip Rose. Rose assessed many companies, including the Best Gas Traction Co. He reported of the Best Company:

> *They are not making much sales effort just now, because their plant is relatively small, and they are so overcrowded with orders that they are simply selling of their reputation. They have no agents, dealers or jobbers, except in foreign countries, because more business comes to them than they can handle. The West looks on a Best Tractor as the automobile public looks at a Packard automobile. . . . I venture to predict that this firm will eventually be the leading tractor manufacturer of the West.*[6]

With the growth of tractor production and the foundry business, the C.L. Best Gas Traction Co.'s Elmhurst location proved inadequate. In 1915, suitable locations for the manufacturing plant were being studied. San Jose offered a ten-acre site, with Richmond and Oakland also seeking the Best Company's relocation.[7] Finally, at the May 6, 1916, board of directors meeting, the announcement was made in favor of the C.L. Best Gas Traction Co. relocating to a site in Melrose.

Enter Daniel Best, described as "San Leandro's grand old man and one of its best boosters."[8] Daniel kept in close touch with the progress of the Gas Traction Co., and with C. L. Since the sale of the Best Manufacturing Company to Holt in 1908, Daniel was officially out of the tractor business. In 1911, he joined A. S. Weaver in incorporating the San Leandro State Bank. The bank allowed Daniel to assist his son in a discreet manor while honoring the terms of the sales agreement signed in 1908. So even though the letter of acceptance was ready to be mailed to the Melrose site land owner, the Southern Pacific Company, the elder Best point blank asked his son what he and his board had to have to bring the manufacturing business to San Leandro. C. L. stated that if the San Leandrites could raise $20,000 toward the purchase of the old Best Manufacturing site (asking price was $30,000), the C.L. Best Gas Traction Co. would move its facilities to San Leandro. That same day, the younger Best was traveling to the East and he agreed to hold the acceptance

---

4. Oscar L. Starr interview, F. Hal Higgins Collection, Special Collections, University of California Library, Davis.
5. C.L. Best Gas Traction Co. Catalog E: 7.
6. Philip Rose, *The Rose Report* (1915): 3.
7. Henry Shaffer, *A Garden Grows in Eden* (San Leandro, CA: San Leandro Historical-Centennial Committee, 1972): 106.
8. "C.L. Best Gas Traction Company Is Coming to San Leandro," *San Leandro Reporter* (June 3, 1916): 1.

In 1911, Daniel Best and A.S. Weaver incorporated the San Leandro State Bank, which was headquartered in the Daniel Best bank building. The Best building still stands and is located at the corner of East 14th St. and Estudillo, in San Leandro, California. Photo Courtesy of Frank Best.

letter until he returned in two weeks.[9] Daniel Best had fourteen days to execute his plan.

Calling on his old friends, A. S. Weaver and C. Q. Rideout, Daniel and the two former directors of the Best Manufacturing Company agreed that it was definitely in the best interests of the City of San Leandro to again have a world-class tractor manufacturer located in the city. To that end, a meeting of businesses and interested citizens was called for Thursday, May 11. As the *San Leandro Reporter* of June 3 related the story:

> *This meeting took place at the city hall and was attended by over three hundred citizens, taxing the hall to its fullest capacity. Over $14,000 was raised that night, with Daniel Best heading the list with $2,500.*[10]

9. Ibid.

10. Ibid.

Over two hundred interested parties present at that meeting signed a pledge certifying their intent to contribute funds to induce the C.L. Best Gas Traction Co. to relocate in San Leandro. One hundred plus individuals were in the $5 to $25 contribution range. The grass roots support to bring good-paying, year-round jobs back to the city was remarkable. Many supporters were skilled workers who had worked for the Best Manufacturing Company in the past and who looked forward to being employed by a Best again.

As Daniel and A. S. Weaver traveled to San Francisco to put up $1,000 as an option on the defunct Best Manufacturing site, now owned by Holt Manufacturing of Stockton, a committee was formed to raise the additional $6,000. With the $20,000 raised and the option obtained on the property, the citizenry awaited the arrival of C. L., scheduled for May 26. Upon his return, Best called a special meeting of his board of directors for June 1 to consider the offer placed on the table by the San Leandro residents.

Daniel Best, Weaver, and Rideout made their presentation. After some discussion, the Best board of directors agreed to move the manufacturing plant and headquarters to San Leandro, providing the city Board of Trustees would allow the building of a crane twenty feet above the street level, thus connecting the buildings that would be separated by Davis Street. With the granting of this privilege by the Board of Trustees, a Best-controlled tractor works was again home in San Leandro.

When transfer of the property title from the Holt Manufacturing Company to the Best people was complete, the removal of the original Best shop buildings commenced on June 28, 1916, by Manning Brothers, a contracting firm out of Fresno, California. Manning Brothers also had the contract to construct the new plant. Plans for the new facility were proudly displayed in the windows of the San Leandro State Bank building. These plans called for the main assembly department to be 981 feet long by 50 feet wide. As more information on the new Best factory became available, the *Oakland Tribune* shared it with the East Bay Area residents:

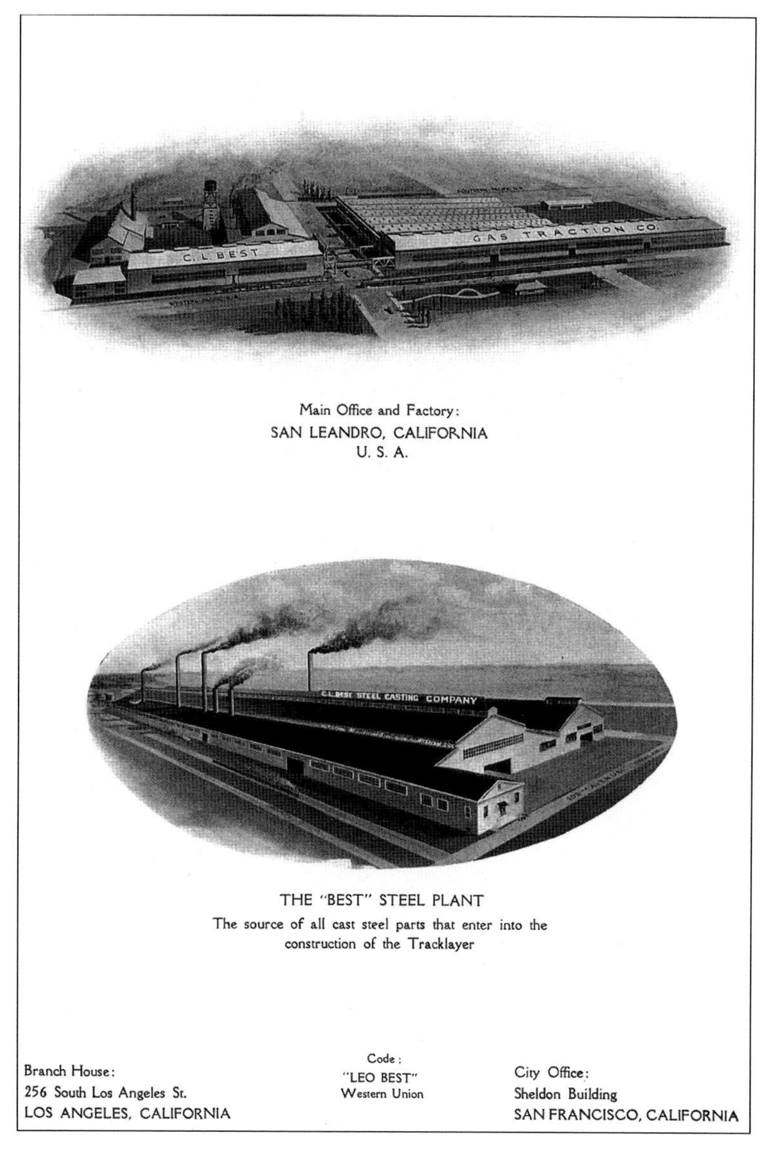

C.L. Best Gas Traction Co. main office and factory located in San Leandro, California. The Best Steel Plant played an important role by providing quality metals for the manufacture of Tracklayers. Once the Best Sixty and Thirty tractors were well into production, custom materials were obtained from the steel mills, making the Best Steel Plant unnecessary. All castings were now done at the San Leandro factory. Company catalog. 1917. Author's collection.

This building is to be of concrete and heavy timber and covered with galvanized iron and 20,000 feet of glass. The floor of this building will be of the same class as macadamized road, approximately 60,000 square feet. The plant will cover three entire blocks and is nearing completion and will be in full operation about October 1.[11]

The construction of the plant was a popular topic in San Leandro during the summer and into the fall season. Old and young were fascinated by the size of the new manufacturing facility. C. L. Best was proud of his new plant and, when approached by a civic committee charged with planning a "housewarming" party and welcome for the relocating firm, offered the use of the main machine shop for the celebration and promised to have a platform erected for the dignitaries and the San Leandro Municipal band.

Friday, October 6, 1916, saw 1,500 townspeople at the new Best plant for the reception and dance. C. L. and Daniel Best were the guests of honor. The San Leandro mayor, Allen Pelton, welcomed the company to the city and spoke of how "pleasing" the relocation was given that Daniel had engaged in business in the same city for many years. The mayor also complimented the citizens on their willingness to raise the money required for the land purchase. C. L. replied to the welcome:

[that] it was the intention of his firm to help in developing the community as far as it lay in its power and that home people would be employed at his shops as much as possible.[12]

The final phase of construction was the installation of machinery. New machinery had been purchased, and some of the equipment from the Elmhurst plant was transferred. The capacity of the San Leandro plant was 25 percent larger than the former plant, with room to double should it be needed. After four months of construction, the first two hundred employees, with a predicted payroll of $17,000 per month, walked into the plant to begin a new era for the C.L. Best Gas Traction Co. and the City of San Leandro. The head of the company also walked, but from his house to the plant. By cutting through the almond orchard and then following the railroad tracks, C. L. had the satisfaction of seeing the newly completed building shining in the morning light.[13] The short walk gave him quiet time to think of the future. But the world was at war, the Holts were in court, and a challenge for the control of his company would test C. L. Best as never before.

11. "Manufacturer's News," *Oakland Tribune* (August 13, 1916): 20.

12. "Towns People at Best Reception," *Oakland Tribune* (October 7, 1916): 5.

13. Personal interview with Daniel G. Best (April 6, 2006).

The Hawkins's Takeover

The years 1915 through 1918 were challenging years for the C.L. Best Gas Traction Co. A patent litigation with the Holt Manufacturing Company began in 1915. The manufacturing plant at Elmhurst was outgrown, and a new facility was constructed at San Leandro. The product line had to be expanded and refined. A distribution network had to be formed. A shortage of raw materials caused by World War I initially slowed tractor production. Financing had to be secured and financial backers kept satisfied. The dissatisfaction of some of the financial backers eventually lead to C. L. Best losing the control of his company.

By displaying its product line at the 1915 Panama Pacific International Exposition, held in San Francisco, the C.L. Best Gas Traction Co. unwittingly fueled the takeover of the company by an eastern investor. Rollin White, founder of White Sewing Machines and White Trucks, from Cleveland, Ohio, saw the experimental 8-16 or "Pony" Tracklayer displayed at the Exposition. Impressed, he purchased it and returned to Cleveland with the intention of using it as a model for a tractor he intended to put into production. Oscar Starr spoke of the sale of the "Pony":

> *We always sold our experimental tractors in those days of short finances. White soon sent C.A. Hawkins, his White Sewing Machine Company sales manager, out to look over the Best and Holt plants with the idea that he might buy one of them and get into crawler tractor manufacture.*[1]

---

1. Oscar Starr interview, F. Hal Higgins Collection, Special Collections, University of California Library, Davis.

C. L. Best driving the 8-16 "Pony" Tracklayer. This photo taken by Daniel Best and sent to his relatives in Iowa. If the "Pony" had gone into production it would have been designated the Model F. Photo courtesy of Frank Best. 1915.

Hawkins was with White and other investors when the Cleveland Plow Company was incorporated in 1916. Their product, a 16-24 horsepower Cleveland Tractor, did not stand up to testing, and Hawkins felt that trying to sell a substandard product was a waste of his time. He resigned from the company, sold his interest to Rollin White, headed west and into a position of power at the Best Company. The *Oakland Tribune* reported:

> *Plans for the investment of more than $500,000 in developing one of the most important machinery industries of the Pacific Coast were made known today. So rapidly have the monetary events of the past few weeks developed that only today was the result of the negotiations known. C. A. Hawkins of San Francisco and Cleveland, Ohio is the capitalist who has paid $500,000 for a share of the Best concern's business.*[2]

Now controlling 51 percent of the company, Hawkins also controlled the board and was elected president, with C. L. Best becoming vice president. With Hawkins in control, the salary of

2. "Best Plant Will Triple Capacity," *Oakland Tribune* (March 18, 1917): 31.

C.L. Best Gas Traction Co. letterhead used during the Hawkins takeover era. Note C.A. Hawkins, President and C.L. Best, Vice-President. 1917. Author's collection.

the president was increased from $6,000 to $30,000 per year.[3] How could a sewing machine salesman from Ohio take control of a tractor manufacturing company in California? What conditions led to the Best Company being vulnerable to a takeover?

When C. L. Best incorporated his C.L. Best Gas Traction Co. in 1910, he had just resigned from the Holt-controlled Best Manufacturing Company and forfeited any money he had in the company. Two-thirds of the money for the new company came from the Woodland, California, area. One-third of the backing was from C. Q. Nelson. Nelson was prominent in banking in the Woodland area and also had a personal tie to Best: He was married to Rose, the sister of Best's wife, Pearl. The other third from the Woodland area consisted of B. G. Peart and other area farmers. The final one-third of the start-up capital was reputed to have come from Daniel Best, but he was not listed in any financial documents.

Investors' hopes were high for the fledgling company. A new production facility was constructed in Elmhurst. Good, solid products were developed, but the costs associated with the research and production were high. In 1914, the company allegedly spent $8,520 to develop a small tractor that saw limited production.[4] Even though the tractors were selling as soon as they were produced, production in 1915 was only about two tractors per week.[5] With little room to expand the Elmhurst plant, a new facility was built on the old Best Manufacturing site in San Leandro. This increased the volume of tractors produced, but

3. "Hawkins-Best Contest Now In Judges Hands," *San Leandro Reporter* (July 13, 1918): 1.

4. Phillip Rose, *The Rose Report* (1915): 3.

5. Ibid.

also incurred costs upward of $50,000 for the land, the buildings, and the necessary machinery.[6]

In February 1915, the Holt Manufacturing Company filed a patent infringement lawsuit against the C.L. Best Gas Traction Co. This case involved the basic track-laying design. The case was complex, involving upward of 1,100 pages of testimony and depositions, many models to visually explain the issues involved, witnesses from as far away as the state of Maine, numerous prior patents that required studying, and a team of attorneys to keep everything progressing. By 1917, legal expenses were high, with no clear end in sight.

The once high expectations of timely, profitable returns on their investments dimmed. The Woodland connections grew dissatisfied. Costs for buildings, product development, and production were significant, and a protracted legal battle loomed. The personal side of the relationship between C. Q. Nelson and his brother-in-law C. L. Best was strained by domestic issues. Best and his wife, Pearl, were having serious problems with their marriage. Dan G. Best reminisced:

> *I remember I was about five years old. Both my parents came to me and sat me down. My dad explained to me that he and my mother weren't getting along, and they wanted to separate and get a divorce. I threw such a fit! They stopped talking about a divorce and stayed together. A lot of the time, when he was home, my dad slept in my room. I remember waking up at night and looking over and there he was—smoking a cigarette. I could see the red tip glowing in the dark.*[7]

When that salesman from Ohio arrived offering cash money for the Best company stock, some of the Woodland investors, led by C. Q. Nelson, accepted the Hawkins's offer. By March 1917, C. A. Hawkins had the leverage he needed to become president and take control of the C.L. Best Gas Traction Co.

Local newspapers published glowing articles about the future plans of the company:

> *Largest tractor contract ever signed—calling for $30,000,000 worth of tracklayer farm tractors—has been awarded by the C.L. Best Gas Traction Company of San Leandro to the Davis Sewing Machine Company of Dayton, Ohio. . . . The contract with the eastern firm was made necessary by the fact that the increasing volume of business can no longer be handled in the San Leandro plant. C.A. Hawkins, president of the C.L. Best Co., is also vice-president of the Dayton concern, and the arrangement was made through his efforts.*[8]

Another news article told of the Best plant in San Leandro tripling its production capacity and that the expansion fueled by eastern capital was "beyond the wildest dreams of the original incorporators."[9] One thing promised by the Ohio expansion was that the Best sales' organization would make an aggressive bid for business in the central, eastern, and southern states when the delivery of the model Forty Tracklayer began in November 1917.[10]

Even though a production plant to the east of the Rockies would cut material costs and increase sales opportunities, not all connected with the Best Company saw the Hawkins moves as progress. H. C. Montgomery, former attorney and general manager for the company, had misgivings about Hawkins. When Hawkins was elected president and was voted the substantial salary increase, Montgomery saw an immediate decrease in his salary as it was based on a percentage of company profits. Montgomery also questioned the ethics of the man when, on a trip through Nebraska together, Montgomery said that Hawkins bragged of foisting three carloads of sewing machines on a dealer

6. "Work to Begin on New Best Plant," *San Leandro Reporter* (June 17, 1916): 1.

7. Personal interview with Daniel G. Best (April 6, 2006).

8. "Big Contract In Tractors Is Awarded," *Oakland Tribune* (May 18, 1917): 8.

9. Ibid.

10. Ibid.

Best Twenty-Five Tracklayer. 1918. Author's collection.

by using high-pressure methods. Per Hawkins, "the dealer went broke, of course, with such bad judgment."[11]

Fred Grimsley was owner of the Best tractor dealership in Stockton. Montgomery gave an account of Grimsley's meeting with Hawkins and Best at a duck dinner held at C. L.'s duck hunting club at Colusa:

> *Grimsley phoned me about Hawkins's plan to get rid of me. Fred had an unerring eye at spotting a four-flusher or phoney proposition, and he had the new Best president sized up before dinner was half over. In fact, this fresh young dealer from Stockton pulled C. L. off to one side as they got up from the table and warned him to get rid of Hawkins before he broke the company!*[12]

11. H. C. Montgomery interview, F. Hal Higgins Collection, Special Collections, University of California Library, Davis.

12. Ibid.

With production slated to begin in Ohio, Oscar Starr, production manager for the Best Company, was sent east by C. L. Best. According to Starr, Hawkins intended to produce the Best twenty-five tractors at the Davis plant. Hawkins wanted to compete with the redesigned tractor being produced by Rollin White's Cletrac company.[13] Starr was not impressed with C. A. Hawkins, viewing him as more a salesman than a serious manufacturer. He wasn't concerned with Hawkins for long. With the United States entering into World War I in April 1917, the Holt Manufacturing Company needed to increase its wartime tractor production at its Peoria, Illinois, plant. Holt's choice for increasing that production? Oscar Starr. When Pliny Holt made him a generous offer, Starr quit the C.L. Best Gas Traction Co. and worked diligently to step up production at Holt's Peoria plant.

Meanwhile, tensions were increasing in San Leandro. The situation between President Hawkins and General Manager Montgomery culminated with Hawkins firing Montgomery. Hawkins was alleged to have accused Montgomery of conspiring with Holt Manufacturing Company attorney C. L. Neumiller to "sell out the C.L. Best Co. to the Holt people for $75,000." This was relative to the patent suit between the Best and Holt companies that was currently in the court system. Hawkins was also quoted as saying in the presence of several people:

> *I have always been suspicious of Montgomery and have suspected treachery on his part. So I had a dictograph installed in his office at the Best Company and had detectives in charge of it. The records show the whole scheme. I am going to discharge him.*[14]

The year 1918 began with C. A. Hawkins as president of the C.L. Best Gas Traction Co. C. L. Best was vice president of the company that carried his name. While Hawkins continued to dabble in corporate affairs, Best tried to conduct business as efficiently as possible: Production at the San Leandro plant had to keep up with the demands of the farmers; an iron workers' strike threatened and finally did idle four hundred workers at the plant; refinements in roller-bearing construction opened the possibility of new tractor designs that would later become the industry standard; restrictions placed on resources by the War Industries Board limited production to a small percentage of the tractors produced in 1917; manufacturing at the Dayton, Ohio, plant had to be planned and initiated without the assistance of Oscar Starr, and the Holt/Best lawsuit continued.

C. L. was aware of the poor opinions the Best dealership owners and staff had of the company president. J. R. Buck, a Hawkins man, came to believe that Hawkins was

> *unscrupulous, untruthful and [a] ruthless individual and that in his [Buck's] opinion, would totally wreck the institution and it would be bankrupt within six months, solely on account of [Hawkins's] wildcat manipulation and his ruinous policies.*[15]

One such proposal involved San Francisco, California, investor Mortimer Fleischhacker. Hawkins had arranged a deal where Fleischhacker would invest $500,000 to finance the reorganization of the company as the Best & Hawkins Tractor Co. When presented with the papers, C. L. Best refused to sign:

> *My father [Daniel Best] may be a bit old-fashioned, but he is opposed to the deal going through, and I want to look into it further.*[16]

Directors Best, Montgomery, and A. S. Weaver were certain that if Hawkins wasn't removed from the presidency, the company that Best founded in 1910 would be done irreparable harm. Hawkins was viewed as unpredictable and, with the power he wielded, could sell the company to the highest bidder. With the Holt/Best lawsuit still in the court system, if offered the chance, would the Holt Manufacturing Company employ the same tactics they used in 1908? That was the year that the Holt Com-

13. Oscar Starr interview, F. Hal Higgins Collection, Special Collections, University of California Library, Davis.

14. "Best Officer Sued For $100,000," *Oakland Tribune* (May 1, 1918): 12.

15. "C.A. Hawkins Is Shown Up By Montgomery," *San Leandro Reporter* (July 6, 1918): 1.

16. "Affidavits Filed In Best Case," *Oakland Tribune* (June 6, 1918): 3.

Best Seventy-Five Tracklayer loaded and waiting on the factory rail spur. Note the wooden crate enclosing the engine. Author's collection.

pany settled its lawsuit with Daniel Best by purchasing the Best Manufacturing Company. C. L. was unwilling to take the chance. He had the Best Sixty tractor in development and didn't want a competitor to get control by default of what he was sure would be the machine that would revolutionize the track-laying tractor industry. So, a board meeting was called for March 13. According to an affidavit later filed with the court, Best is said to have called the meeting to consider repairs that Hawkins deemed unnecessary. Hawkins and board member Oscar T. Barber boycotted the meeting. J. R. Buck, the seventh board member, received a request from President Hawkins to also stay away from the meeting so as not to constitute a quorum, but Buck did attend the meeting.

To take back control of the board, Best had to get control of more of the company's stock. At that time, Hawkins had the majority at 1,369.5 shares with Best having 1,000 shares out of a total of 2,447.5 shares. Nine hundred shares remained in the company's treasury.[17] H. C. Montgomery had retained his permit to sell stock, so the first order of business was to sell those nine hundred shares in the treasury to Daniel Best. The Bests now

17. Ibid.

had control of 1,900 shares, and the second order of business was to reinstate C. L. Best as president of the company. Montgomery became vice president and secretary with J. R. Buck treasurer.

The deposed Hawkins alleged the act was a conspiracy to deprive him control of the company and on April 3 filed suit against Best, Montgomery, Weaver, and Buck. The complaint claimed that the issue of the stock was illegal. The complaint further claimed that Buck supported the stock sale and subsequent reinstatement of Best as president because of promises of monetary gain. As reported by the *Oakland Tribune* on May 4:

> *Hawkins contested the entire transaction, which resulted in his being unseated as president, and carried the matter into the courts in a suit to annul the sale of the stock to Daniel Best.*[18]

Hawkins's next move was to request and be granted a court order preventing the Best Company from entering into any long-term contracts while the court case was pending.[19]

C. A. Hawkins was also on the receiving end of a lawsuit. When completing his deposition for the conspiracy case, Hawkins angered Montgomery when the topic of Montgomery's firing was addressed. Hawkins denied making the statement about improper dealings between Montgomery and Neumiller. Montgomery immediately brought a $100,000 slander suit against Hawkins.[20] The *San Leandro Observer* reported that in early August, Hawkins, still a stockholder even though no longer the president of the company, arrived at the corporate offices and got into a spirited discussion with Montgomery. Fists flew and the two combatants had to be separated by others in the office. Hawkins left San Leandro and returned to his home in San Francisco to await Superior Court Judge W. S. Wells's decision on the illegal stock sale case.[21]

He didn't have long to wait. On August 10, 1918, Judge Wells dissolved the restraining order and refused to issue an injunction requested by Hawkins concerning the stock sale that led to C. L. Best regaining control of his company. The court's decision allowed Best to terminate the contract with the Davis Sewing Machine Company. Production east of the Rockies would have to wait. All Best Tracklayers would be manufactured at the San Leandro plant.

The purchase of Hawkins's shares of Best stock in December 1918 brought the Hawkins's era at the C.L. Best Gas Traction Co. to an end. Hawkins reportedly received $400,000 for those 1,370 shares and officially retired. As reported in the *Oakland Tribune*:

> *The capital required for the purchase of the Hawkins interest is being furnished by Best without a bond issue, it is stated, and the deal will remove the friction that has handicapped the operation of the plant and enable it to go ahead with its program of expansion.*[22]

This settlement ended one court battle for C. L. Best, but another ongoing litigation consumed much of the president's time and the company's finances.

18. "Litigation Over The Best Plant," *Oakland Tribune* (May 4, 1918): 17.

19. "Restraining Order Bars Long Contracts," *Oakland Tribune* (April 20, 1918): 5.

20. "Best Co. Officer Sued For $100,000," *Oakland Tribune* (May 1, 1918): 12.

21. "Hawkins Was Aggressor," *The San Leandro Observer* (August 10, 1918): 1.

22. "Gas Tractor Control Assumed Best," *Oakland Tribune* (December 1, 1918): 1.

No. 874,008. PATENTED DEC. 17, 1907.

B. HOLT.

TRACTION ENGINE.

APPLICATION FILED FEB. 9, 1907.

7 SHEETS—SHEET 1.

Fig. 1.

WITNESSES:

INVENTOR

Benjamin Holt

BY

Geo. H. Strong.

ATTORNEY

Benjamin Holt Traction Engine. Patent number 874,008. Patented December 17, 1907.

Best and Holt in the Courts—Again

Benjamin Holt had seen what lengthy court battles could consume in both time and money. The 1904 copyright infringement suit filed by the Best Manufacturing Company against the Holt Manufacturing Company had dragged on until 1908, when the Holt firm acquired the Best Company. Now faced with competition from the upstart C.L. Best Gas Traction Co., Holt still felt he had no choice but to resort to litigation.

This new court case, brought to the San Francisco Superior Court in January 1913, involved the use of the Best name. C. L. pointedly used the fact that he was the only Best in the business in his advertising. Given the proximity of the Elmhurst-based C.L. Best Gas Traction Co. and the Holt-owned San Leandro–based Best Manufacturing Company, as well as the name similarity, there was considerable confusion of mail, telephone, and wire orders.

C. L. retained H. C. Montgomery as his attorney. Montgomery had been the junior partner to J. J. Scrivner, who served as Daniel Best's counsel in the earlier patent infringement suit. As told by Montgomery:

> *Holt hired Townsend, one of San Francisco's big names in patent law. He had 400 pages of affidavits. I defended C.L. on the grounds he had something entirely new, as his firm name indicated—gas. Holt still had a big steam outfit; Best, a small gasoline rig. We licked 'em. It cost Holt a lot.*[1]

---

1. F. Hal Higgins Collection, Special Collections, University of California Library, Davis.

A standard-model Lombard steamer is shown pulling sleds loaded with logs in northern Maine. The man in the fur coat at the rear of the first sled is Alvin O. Lombard. Photo courtesy of the William B. Lynch Collection.

The suit of 1913 was not the end of the Holt and Best concerns meeting in court. A new suit, No. 167, filed by the Holt Manufacturing Company with the courts on February 19, 1915, involved patent infringement. The C.L. Best Gas Traction Co. was the defendant. The patent cited was No. 874,008, issued to Benjamin Holt on December 17, 1907, and was for

> *certain improvements and discoveries in traction engines of the self-laying track type. That said patent was issued for a new and useful invention never before known or used, or in commercial use before the invention thereof by said Benjamin Holt.*[2]

Benjamin Holt had sold and assigned all of his patent rights and titles to the Holt Manufacturing Company.

This infringement case was complex, with Best and Holt providing much documentation. A large share of that documentation involved early self-laying track as it pertained to tractors. Both C. L. Best and Benjamin Holt had researched the earlier self-laying track patents of other inventors. Most notable was Alvin O. Lombard of Waterville, Maine. Lombard was credited with being the first manufacturer to build and sell a track-laying machine that was a commercial success. It was estimated that he had sold almost two hundred of these track machines from 1902 through 1915. He had also sold a license to produce and sell this type of machine to the Phoenix Manufacturing Company of Wisconsin. This company agreed to pay Lombard the sum of $12,000 annually on the first day of January for eleven years until the full amount of the contract, $132,000, was paid. It was further stipulated that should the Phoenix Company build

2. U.S District Court documentation (February 19, 1915).

more than twelve machines per year, an additional $1,000 per machine built would be assessed. The Phoenix territory was to include all lands west of Pittsburgh, Pennsylvania.[3] Lombard's first patent for a track-type engine was dated May 21, 1901, over six years earlier than the Holt patent in question. Lombard was also granted a second patent in 1907 for an improved engine.

Benjamin Holt was aware of the 1901 and 1907 Lombard patents, but was informed by his legal counsel that his track-type machine did not infringe on the 1901 patent because of how the engine was carried: roller-belts for the Lombard engine and small wheels for the Holt machine. The 1907 Lombard patent supposedly did not include features consistent with the Holt tractor and would pose no problem. Once again Holt turned to legal counsel, this time about the value of purchasing the Lombard patents. He was informed that purchasing the Lombard patents would be of little value to the Holt Company. And furthermore, many of the supposed "novel" features of both companies' patents could be found in numerous earlier patents held by various inventors. While Ben Holt would not be able to claim a monopoly on the manufacture of track-type engines, he was confident that neither could he be charged with infringement of the Lombard patents.

C. L. Best also turned to legal counsel about both the Lombard and Holt track-type tractor patents. While still employed by his father at the Best Manufacturing Company, the younger Best sought an opinion from E. L. Thurston. Attorney Thurston studied the patents in question and determined that neither could be considered a pioneer patent as both had drawn on prior art.[4]

In 1907, when C. L. received Thurston's patent opinion, the Best Company was still Best controlled; it was in a lengthy legal battle with the Holt Manufacturing Company about Holt infringement of a Daniel Best patent, and it was building large round-wheel steam- and gas-traction engines. Drawing on his long experience in the traction engine business, C. L. found

> *that the large round drive wheel tractors were incapable of good traction in unfavorable conditions; the wheels slipped, mired and could not pull efficiently in rainy weather, or in soft sandy or marshy soil; that through trade-journals, scientific periodicals and otherwise, [he] long since learned of the existence of what was known as the "walking wheel," "ped rails," "platform wheels" and "track layer tractors."*[5]

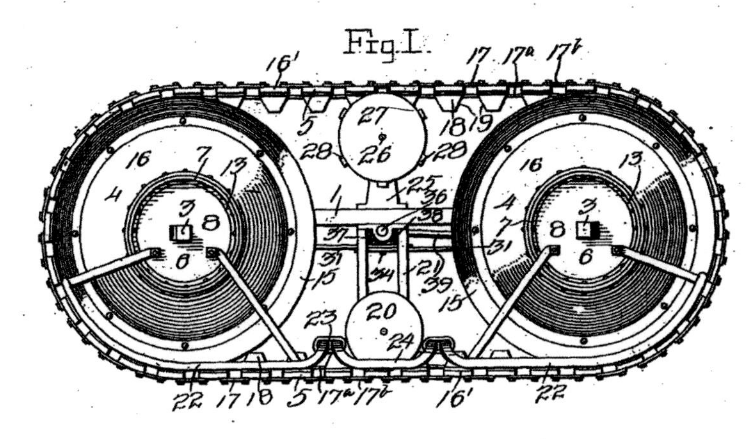

Harvey Beckwith Traction Engine. Patent number 754,409. Patented March 15, 1904.

Best continued to ponder possible practical solutions involving the platform wheel design. Harvey Beckwith, another East Bay area inventor, patented his concept of a "self-laying track type" device in 1904. The Beckwith design, while never going into production, provided Best with a local example of a variation on the platform wheel design.

With the purchase of the Best Manufacturing Company by Holt, C. L. would have been strategically placed to contribute to the refinement of the existing Holt track-type tractor. When events led to Best leaving the Holt-controlled company, he established his new company in Elmhurst and vigorously pursued the research and development of his track-layer tractor.

3. "The Lombard Log Hauler," *Daily Kennebec Journal* (February 19, 1907): 3.

4. Pliny Holt papers, The Haggin Museum, Stockton, California.

5. C. L. Best deposition, U.S. District Court-Northern District of California (August 31, 1915).

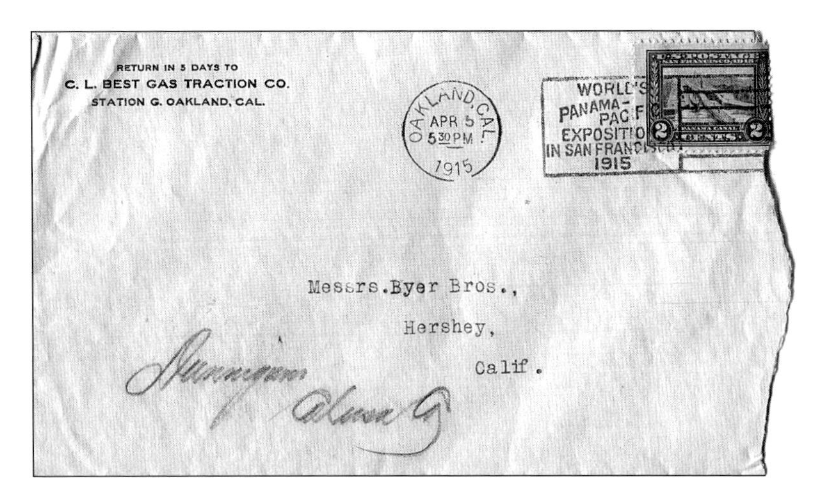

Envelope from the C.L. Best Gas Traction Co. to the Byer Bros., Hershey, California, postmarked April 5, 1915. Note the Panama-Pacific Exposition cancellation. Byer Bros. owned a 70 H.P. Best Tracklayer. Author's collection.

The San Francisco World Fair, officially known as the 1915 Panama Pacific International Exposition, served as a backdrop for legal maneuvering by the Best Company attorney, H. C. Montgomery. The Exposition ran from February 20, 1915, through December 4, 1915. Many farm tractor and implement manufacturers set up displays in the Palace of Agriculture, including the Best and Holt firms. The Holt Manufacturing Company spared no expense in setting up its nearly 15,000-square-foot display. Its entire line of machines was shown. It was acknowledged in the press that this was one of the finest exhibits of farm machinery ever seen at a world's fair.

The C.L. Best Gas Traction Co. exhibit was modest by comparison, only showing three production tractors and an experimental tractor: a 75 H.P. Tracklayer, a 30 H.P. Tracklayer for orchard work, a round-wheel tractor, and the experimental 8-16 "Pony" crawler. However, this modest display seemed to be the tipping point for Ben Holt, and the patent infringement lawsuit against C. L. Best was filed on February 19, one day before the exposition opened.

The 1915 Exposition was used as a lure by Montgomery to get patent holder Alvin Lombard to leave his home in Waterville, Maine, and travel to the West Coast. While studying the prior patents involved in the infringement lawsuit, Montgomery concluded that Lombard's patents would be key to the defense receiving a favorable outcome to the lawsuit. When written correspondence to Lombard went unanswered, Montgomery boarded a train east and met with Lombard to tell him of the situation. This visit served to rekindle the ill feeling held by Lombard about Benjamin Holt. When introduced to Montgomery, Lombard exclaimed:

> *By God, young man, I'm glad to meet you. If God Almighty would charter me to kill a man, I'd get on a train, go out to California, and kill old Ben Holt.*[6]

According to Lombard, he had traveled to California in 1910 to discuss his patent rights with Ben Holt. Having met with Holt, Lombard felt he had gotten snubbed:

> *I remember... Mr. Holt saying to me, "Mr. Lombard, why don't you keep in your field up there and in the lumber woods, and we keep in ours out here, and there is chance enough for both to work." That was simply a proposition that Mr. Holt made to me as long as I was laying it to him, his infringing, and using my device.... The last time I saw him he says—he just says, "Mr. Lombard, we can't tell what to do, right today" to that effect, and for me to go home and he would write us, and thought we ought to come to some understanding, so that I could work in my territory and he work in his and everything would be all right and, of course, I knew I wouldn't hear nothing from him. I knew he was just doing that to get rid of me, but anyway that was all I could do at that time particularly, so I went home and never heard from him.*[7]

Realizing the value that Lombard's testimony could have for his case, Montgomery invited Lombard and his daughter, Grace Lombard Vose, to travel west to take in the 1915 Exposition—at

6. H. C. Montgomery interview, F. Hal Higgins Collection, Special Collections, University of California Library, Davis.

7. A. O. Lombard deposition, U.S. District Court, San Francisco, California (December 9, 1915).

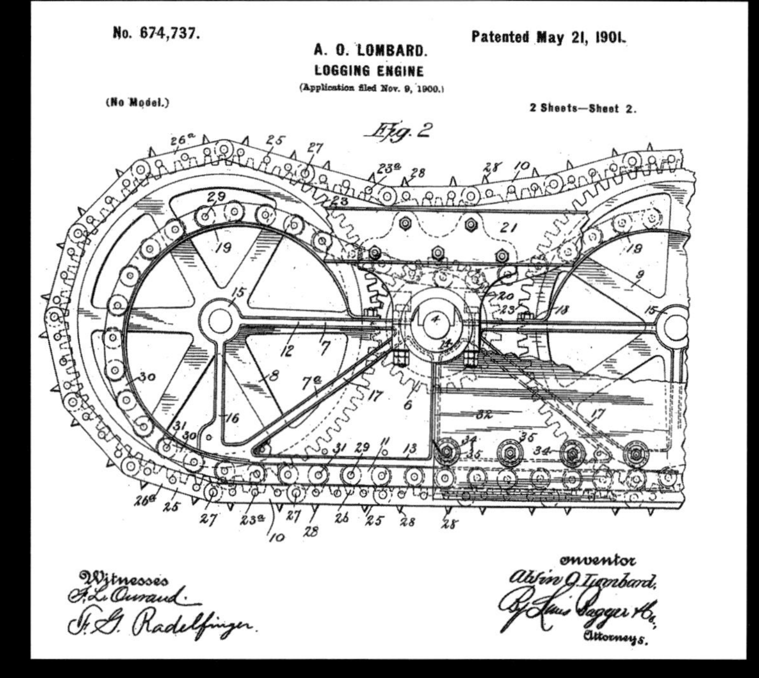

A.O. Lombard Logging Engine. Patent number 674,737. Patented May 21, 1901.

Close-up photo of Lombard track details. Photo courtesy of the William B. Lynch Collection.

the expense of the C.L. Best Gas Traction Co. Of course, the journey would also include giving depositions to the District Court as witnesses for the defense.

Both A. O. Lombard and his daughter did give depositions about the manufacture and distribution of the patented Lombard log haulers. When asked why he hadn't pursued Holt in court, Lombard said given what it would have cost him in both time and money, and that the suit would have to be filed in California, in 1910 he didn't have the resources or the available free time to bring such a suit. But by 1915, he had handed the day-to-day manufacturing operations to his brother and son-in-law and was financially able to consider litigation.

Henry Montgomery had also traced the whereabouts of several of the Phoenix-licensed machines to the Western Lumber Company in Montana. In his deposition, Charles H. Richardson, manager of the Western Lumber Company, stated that in March 1906, Ben C. Holt, accompanied by his Uncle Benjamin Holt, arrived in Missoula, Montana, and inspected the Lombard/Phoenix engine. The two Holt men then spent the night with Richardson and his family.[8]

Both the Pan Pacific Exposition and the year 1915 were coming to an end, but the legal battle between the two companies showed no sign of closure. After studying the patent laws involved with the Holt/Best case and hearing the various depositions, H. C. Montgomery presented his findings to C. L. Best and recommended a striking course of action: Purchase the two A. O. Lombard patents and, because those patents predated the Holt patent, sue the Holt Manufacturing Company for the infringement of the Lombard patents. While previous counsels for both litigants saw little value in owning the early patents, Montgomery believed those same patents were the key to a successful outcome. C. L. Best agreed and Montgomery went back to Maine and purchased the two patents for $25,000.

Because Lombard had a royalty contract with the Phoenix Manufacturing Company, Montgomery and Lombard traveled to Eau Claire, Wisconsin, to obtain cancellation of the 1907 contract. This eliminated any contention concerning manufacturing rights and geographic territory controlled by the Wisconsin company. The stage was set; the next move belonged to Best's attorney, H. C. Montgomery. From an article in the *Oakland Tribune*:

> $3,000,000 SUIT RENEWS TRACTOR WAR
> *With the filing of a suit for $3,000,000 damages in the United States District Court, the C.L. Best Gas Traction Co. of Oakland has renewed efforts to recover from the Holt Manufacturing Company of Stockton on patents covering tractors which have been manufactured by the latter concern since 1909. This action involves the ownership of patents issued in 1901 to A.O. Lombard of Maine.*[9]

The Best legal team based the $3,000,000 value of the suit on the estimate of at least three thousand Holt tractors being sold since 1909, with a minimum profit of $1,000 per tractor.[10] According to the Bill of Complaint, the Lombard patents, now Best owned

> *for the first time conclusively demonstrated that the track-laying traction engine problem had been solved and said invention went into extensive commercial use.*[11]

The filing of this suit, No. 240, put the Holt Company on the defensive. It would have to show how the Lombard patents were not novel and did not represent a commercial success. Now, the Lombard/Best patents could not be ignored so the extent of the trial broadened considerably. The additional testimony and affidavits required would span the nation—from Maine to Wisconsin to California. This case, No. 240, was to be tried during the July 1916 term for the District Court of Northern California. However, the defendant, Holt Manufacturing, had not filed

8. Charles H. Richardson deposition, U.S. District Court, San Francisco, California (March 6, 1916).

9. "$3,000,000 Suit Renews Tractor War," *Oakland Tribune* (December 24, 1915): 5.

10. Bill of Complaint, U.S. District Court, filed December 23, 1915.

11. Ibid.

its expert affidavits in the requested time frame and had instead received extensions for that filing.[12]

Best attorneys, J. J. Scrivner and H. C. Montgomery, filed a motion with the District Court on June 13, 1916, requesting the consolidation of the *Holt v. Best* case, No. 167, and the *Best v. Holt* case, No. 240. The reasons for requesting the consolidation were that both cases would require thorough review of the Holt and Lombard/Best patents; that both parties had prepared models, diagrams, photographs, and drawings that would be used in both cases; that much testimony would be applicable to both cases; and that both cases were represented by the same counsel. The Best legal team believed that

> *the time and convenience of the Court will be greatly conserved by consolidating cases No. 167 and No. 240 and litigants saved many thousands of dollars in expenses.*[13]

The Court did grant the consolidation motion. Both litigants continued to gather information through the remainder of 1916 and into 1917.

It was on February 24, 1917, the Best counsel filed a motion to refer the consolidated case to a Master in Chancery:

> *The Court may direct a reference to the Standing Master of any cause in equity or any issue therein presenting matters of complicated detail, or of a technical or scientific character, or wherein the hearing is likely to be prolonged to such an extent as to interfere seriously with the ordinary business of the Court.*[14]

The Court was presented with an accounting of just how complicated the case would become. The three separate patents under scrutiny contained over sixty claims and large numbers of drawings and specifications. Because of the complicated nature of the machines, study would be required by the court before it could fully understand the testimony offered by the parties. At the time the motion was made, vast amounts of information had already been introduced: 549 pages of printed testimony; 581 pages of legal-size typewritten pages of testimony and depositions; 30 to 40 elaborate models of various mechanisms involved in the Lombard, Holt, and prior patents; 100 printed patents already partially analyzed with perhaps 100 more to be introduced concerning prior art; and numerous drawings, sketches, and photographs. If the case was to be tried in open court, upward of sixty to seventy witnesses could be called for the trial, with many coming from Maine and Wisconsin. It was estimated that it would require nearly six months of study to fully comprehend the issues presented. On July 30, 1917, Judge William C. Van Fleet granted the motion and referred the case to Harry M. Wright, Standing Master in Chancery of the Court, to take testimony and report back to the Court his findings and conclusions.[15]

The patent litigation continued to consume much time and money from both companies, and the court case wasn't the only item of concern for the Best and Holt companies. The Holt Manufacturing Company was heavily invested in tractor production for World War I. Its domestic production slowed to a trickle. When the war contracts were fulfilled and the war was winding down, Holt no longer had an advantage in the domestic market.

The C.L. Best Gas Traction Co. attempted to get an Army contract by producing a small tractor intended to move a three-inch gun. This tractor was shipped to Rock Island, Illinois, for a field trial. Ralph Easton, a Best employee and tractor demonstrator, was there:

> *During the war, I was sent to Rock Island, Illinois.... We [Best] had built in the tool room at the San Leandro plant a high-speed tractor. This machine was designed by Mr. Starr and Mr. Best, and was put out in less than six weeks time.... We made 20 mph with a full three inch gun, which included four ammunitions carts, a limber and*

12. Motion to Consolidate, filed June 13, 1916.

13. Ibid.

14. Motion to Refer to Master in Chancery, filed February 24, 1917.

15. District Court, Northern District of California (July 30, 1917).

a gun. It was a mystery to me why this machine was not adopted, as it showed the best report of anything that was tested at the arsenal at the time.[16]

Holt Manufacturing Company Vice President Murray Baker played a pivotal role in the contest for war contracts. Baker had a cousin who worked at the War Office in Washington, and Baker told of a visit the two cousins enjoyed:

I dropped into the War Office... and I noticed the Best men waiting outside the door. My cousin asked me how he should handle them. I merely reminded him that he was a busy man. He should see them when he had the time. And I winked. My cousin kept those Best boys out there cooling their heels.[17]

Having been shut out of war contracts, the Best Company concentrated on improving its products and expanding its markets. Best had received assurances from Washington that he would receive the materials needed to allow him to continue to build for the domestic farm market. But there were challenges for C. L. Best: A corporate takeover left him with little control of the company he founded and the finances for the company were tight. The lawsuit with Holt consumed an incredible amount of time and energy, and the litigation wasn't confined to just the office. Per Daniel G. Best, C. L.'s son:

When I was a little kid sitting at the dinner table, the talk always went back to Holt. It was always sue, sue, sue. That's all it seemed they ever talked about.[18]

With the control of the Best firm still in litigation, the Holt Company was unsure of the outcome and how it would affect the patent case. As costs and outside concerns mounted through

"Manufactured by C.L. Best Tractor Co. San Leandro, California. Licensed under United States Patents of the Holt Manufacturing Company USA." A brass plate was affixed to all Tracklayers manufactured after the settlement of the Holt v Best lawsuit. Author's collection.

1917 and into 1918, the pressure to negotiate a settlement was strong. Any challenge by C. A. Hawkins to the C.L. Best Gas Traction Co. was eliminated in December with the purchase of the Hawkins's stock by the Best concern. The original litigants would be the ones to settle the case. A compromise settlement was reached, and the final decree signed on December 30, 1918, ending nearly four years of litigation. The final decree did state that Benjamin Holt's patent, No. 874,008, was valid and that Holt was

the first and original inventor of the subject matter of the claims of said patent; that the Holt Manufacturing Company is the owner of said Letters of Patent; that the Defendants, C.L. Best Gas Traction Co. & C. L. Best have infringed on 12 claims of said Letters of Patent. Appearing that the parties have entered into a License Agreement under said Letters of Patent, it is ordered, adjudged and decreed that no injunction issue during the life of the said License.... that the Plaintiff recover of the Defendants nothing as profits and damages... and that neither of the parties recover costs against the other.[19]

16. Ralph Easton interview (1931), F. Hal Higgins Collection, Special Collections, University of California Library, Davis.

17. Murray Baker interview (1952), F. Hal Higgins Collection, Special Collections, University of California Library, Davis.

18. Personal interview with Daniel G. Best (April 6, 2006).

19. Final Decree: In Equity No. 167, U.S. District Court, Northern California, Second Division (December 30, 1918).

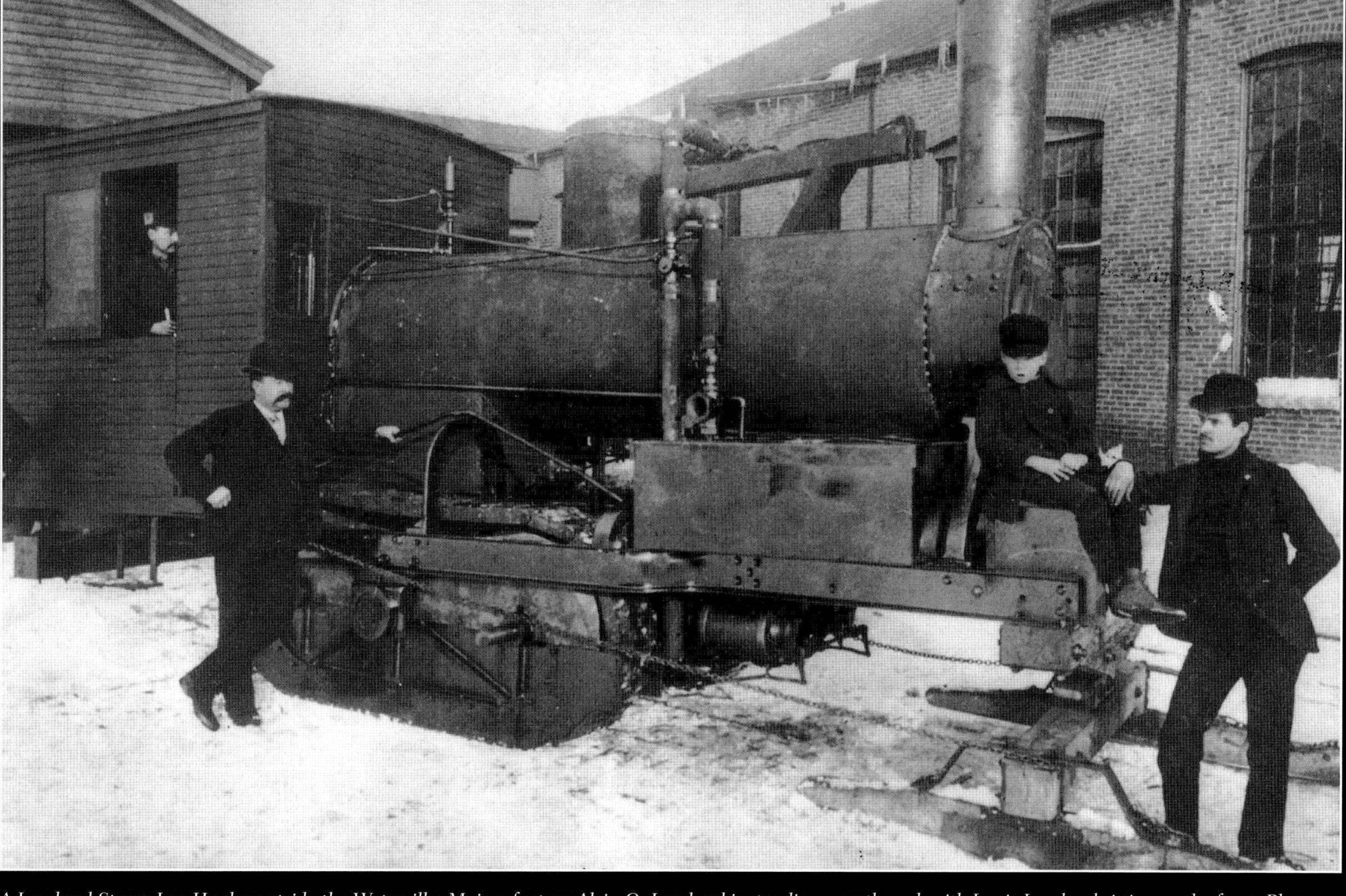

A Lombard Steam Log Hauler outside the Waterville, Maine, factory. Alvin O. Lombard is standing near the cab with Louis Lombard sitting on the front. Photo courtesy of the William B. Lynch Collection.

Taken at face value, the Holt Manufacturing Company did win its patent suit against C. L. Best, but there was much maneuvering behind the scenes. Henry C. Montgomery was the negotiator for the Best Company. During the final phase of negotiations, C. L. Best commented to Montgomery that he would be satisfied with $50,000 and a licensing agreement that allowed him to continue to manufacture the Tracklayer. Montgomery spent four days at Stockton, location of the Holt Manufacturing Company. When he returned to San Leandro, he brought excellent news for C. L. Best. Montgomery did indeed get a licensing agreement from Holt, but the price for the patent rights was four times the amount for which Best had hoped: $200,000 cash.[20] All future crawler tractors manufactured by the C.L. Best Gas Traction Co. would bear a plate with the words "Licensed under United States Patents of The Holt Manufacturing Company U.S.A."

The Holt Manufacturing Company now had the responsibility of defending its patents against other manufacturers who were allegedly infringing on those patents. The company had to be vigilant as many saw the value of a track-type tractor. To combat the infringements, Holt hired former Best attorney H. C. Montgomery. It was Montgomery's responsibility to demand licensing agreements from the infringing companies. Fees ranged from $25 to $100 per tractor. Owning and defending the patents was an ongoing expense for Holt. Even with an estimated $10 million in assets, the long litigation with Best was a financial drain on the company. Estimates place the costs at over $500,000.[21]

The settlement of the patent case was also noted by newspapers in A. O. Lombard's home state of Maine. As reported by the *Bangor Daily News*:

WATERVILLE MAN IS PROVED ORIGINATOR OF 'CATAPILLAR [SIC] TREAD'

At a cost of a million dollars, the claim of A.O. Lombard to being the originator of the famous catapillar tread of war tanks and traction engines, with Waterville as its birthplace, has been established. Mr. Lombard's ownership of the basic patent has been proven and henceforth all right to manufacture must come from him or those to whom he may delegate authority.

Seven years ago, Mr. Lombard sent a log hauler to Lothrop Montana for lumbering operations. Representatives of the Benjamin Holt Co. of Stockton Cal. heard of the new tractor and sent men to investigate, and soon began to manufacture a similar machine. Then the Best Manufacturing Co. of Oakland Cal. began making according to its own ideas. The Holt concern sued for infringement, claiming to be pioneers in the business.

This started the Best people looking up the history of the invention and they found that Mr. Lombard had a clear title to it and made arrangement with him to fight the Holt concern. So, he stepped back and let the two powerful corporations fight it out.

By the time the case was ready for a real trial, both sides, ready for a compromise, got together. They now have settled with Mr. Lombard. He retains the manufacturing rights to New England and Canada and equal right with the Best and Holt people and the New York concern to which he sold right to manufacture. The Best, Holt and Lombard Co. of Waterville and the New York concern [Linn] now hold absolute control of the catapillar [sic] traction engine principally.[22]

C. L. Best was now done with the legal wranglings and could devote his efforts to his product line. The purchase of the Hawkins block of stock secured control of the company. The $200,000 received from the sale of the patents put the company on firm financial footing. With the new plant in San Leandro, the threat of litigation behind him, money in the bank, and an innovative tractor ready for production, the year 1919 began with the C.L. Best Gas Traction Co. poised to assume a leadership role in the track-type tractor market.

20. H. C. Montgomery interview, F. Hal Higgins Collection, Special Collections, University of California Library, Davis.

21. Reynold M. Wik, *Benjamin Holt & Caterpillar: Tracks & Combines* (St. Joseph, MI: American Society of Agricultural Engineers, 1984): 112.

22. "Waterville Man Is Proved Originator of 'Catapillar' Tread," *Bangor Daily News* as reprinted by the *Daily Kennebec Journal* (November 8, 1918): 4.

Why Tracklayers Are Best

The Best Tractors of today are the result of long years of observation in factory and field. The accumulation of refinements and the improvement of design have made of the machine an efficient, dependable power plant able to conquer the undeniably severe conditions under which track-type tractors are called upon to work.[1]

The path that culminated with the above statement began in 1910, when the C.L. Best Gas Traction Co. was incorporated. With its production facility located in Elmhurst, California, the Best Company developed and produced the Best 70 H.P. and 75 H.P. Tracklayers, the "Humpback" 30 H.P. Tracklayer, the 8-16 "Pony," as well as the 30 H.P. and 40 H.P. Tracklayers. The path then led to the new facility in San Leandro. The Best 25 H.P. Tracklayer with antifriction bearings was in the design stage at San Leandro in 1917. The year 1917 also saw the path nearly obliterated when the balance of power shifted from C. L. Best to C. A. Hawkins. But that turn of the path also led to Best immersing himself into designing the Best Model Sixty, which was destined to become his signature tractor.

The Model 70 H.P. tractor served as Best's entry into the track-laying tractor field. With an increase in engine RPM, the tractor horsepower rating was increased to seventy-five and sold as the 75 H.P. Tracklayer. Two prominent features of the Model 75 tractor involved the tracks. The link pins that

1. C.L. Best Tractor Co. Catalog (1922).

C.L. Best Gas Traction Co. Model 75 Tracklayer. Company catalog. 1917. Author's collection.

Best's Model 90 Tracklayer. Note that on the frame the words Track and Best are painted but the word Layer is merely sketched in. Author's collection.

provided the pivoting points of the tracks were stumbling blocks for manufacturers. These pins were exposed to a tremendous amount of friction. Reducing this friction would result in less upkeep and maintenance costs and yield more usable drawbar horsepower. Best's answer to the problem was his patented frictionless rocker joint link. The second important feature of the track was the oscillating roller frame. Unique to tractors of this time period, the Best Model 75's hinged track unit allowed the tractor to maintain its power on uneven surfaces.

Another feature, first employed on the Model 75 tractor, was the use of cut and machined steel transmission gears. These were enclosed and ran in oil. The use of steel in the gearing, while expensive, added longevity. By using the enclosed transmission case, Best significantly reduced wear to the gears by eliminating exposure to dirt. Running the gears in an oil bath provided constant lubrication and required no effort from the operator. Based on his design for his 80 H.P. round-wheel tractor, Best continued to use the front tiller wheel for steering and a differential to compensate for when the tractor executed a turn. Only now that tractor had two rear track units instead of two round wheels.

The larger 90 H.P. Tracklayer, while similar to the Model 75, was built for very heavy workloads. The Model 90 excelled in hauling loads in difficult conditions. The use of the side friction drum clutches allowed the tractor to be easily handled and turned on a short radius. While marshy river-bottom lands and soft, sandy desert soils could be problematic to controlling the slippage of differential-controlled tracks, they presented little problem for the Model 90.

Not all applications required the size and power of the Model 75 or the Model 90 tractors. Best built the "Humpback" 30 H.P. tracklayer with smaller operations in mind. The term *humpback* referred to the tractor's unique elevated drive sprockets. This tractor continued the tiller wheel for steering.

The Elmhurst-built 8-16 (8 H.P. drawbar and 16 H.P. belt) "Pony" Tracklayer did not employ a tiller wheel for steering. While it never was a production model, it would nevertheless play an important role in the C.L Best Gas Traction Co.'s future.

Model 30 H.P. and Model 40 H.P. Tracklayers were the first production Tracklayers without the front tiller wheel. This feature, when combined with the expanding shoe-steering clutches, allowed the tractors to turn within a fifteen-foot circle.

Best 30 "Humpback" tractor doing orchard work. Note the elevated sprocket at the rear. This design feature would appear again in 1978 with the Caterpillar D10 tractor. Special Collections, University of California Library, Davis.

8-16 H.P. "Pony" Tracklayer. The "Pony" was Best's first tractor without a front tiller wheel. This model was featured at the 1915 Panama Pacific Exposition held in San Francisco. Company catalog. 1915. Author's collection.

Proven features from the Model 75 tractor incorporated into the design of the Model 30 and Model 40 tractors were the oscillating roller frame; the frictionless rocker joint; cut and machined steel transmission gears, enclosed and running in oil; and the use of alloyed metals whenever necessary.

The first tractor designed and built at the new San Leandro plant was the Best Model 25 H.P. Tracklayer. Intended for orchard use and for moderate-sized farms, the Model 25 incorporated significant modern engineering. It was advertised as being able to pull a plow of four 10" bottoms at a depth of 7". This tractor also featured two important Best track features: the oscillating roller frame and the patented rocker joint.

With the Model 25, Best was one of the first tractor manufacturers to embrace the use of antifriction bearings. Most tractor manufacturers used plain bearings. Best explained the role a bearing played and why the use of antifriction bearings was advantageous to his tractors:

> *A bearing permits the parts which it supports to revolve freely, and at the same time holds them firmly, so as to prevent lost motion or looseness. Anti-friction bearings differ from plain bearings in that they substitute a rolling movement for a sliding or scraping movement. This reduces friction to a minimum and, as friction absorbs power and causes*

wear, these bearings not only prolong the life of the tractor, but save in fuel and deliver more power to the drawbar.[2]

The design of the tractor dictated the quantity and types of bearings required. The Model 25 contained thirty-five antifriction Hyatt and Timken brand bearings. C. L. Best's use of Timken antifriction bearings predated Henry Ford's use of such bearings on his famous Model T.[3]

Still another use of modern technology in the design of the Model 25 Tracklayer was multiple disk steering clutches. New to this tractor, these clutches replaced the expanding shoe type clutches used on the Model 30 and Model 40 Tracklayers. Disengaging one steering clutch allowed the corresponding track to slow or stop, while the other track continued to work under full power, facilitating a turn. While more expensive to produce, the multiple disk steering clutch required less maintenance during the life of the tractor. C. L. Best firmly believed that the value of his products was more than purchase price. This belief was stated in many company catalogs:

(Top left) *Best Model Forty Tracklayer. Originally introduced at Elmhurst as the Forty-Five Tracklayer. The 45 horsepower designation was derived from the S.A.E. formula for calculating horsepower, but it was soon discovered that this model could only produce 40 horsepower so the name was changed.* Author's collection. (Top right) *Charles (Tex) Manning driving the first Model Twenty-Five tractor. Manning was C. L. Best's brother-in-law.* Photo courtesy of Dan G. Best II. (Bottom right) *C. L. Best's Model Sixty Tracklayer. At serial number (S/N) 1126A, the design shown became the standard configuration for the Sixty tractor. The last major change was at S/N 2201A when the engine went from splash lubrication to forced lubrication.* Company catalog. Author's collection.

2. "Why Best Tractors Are Tracklayers," C.L. Best Gas Traction Co. (1920): 5.

3. Bettye H. Pruitt, *Timken: From Missouri to Mars* (Boston: Harvard Business School Press, 1998): 58.

In buying a machine of this type—whose value lies entirely in its utility—the question is not so much "what does it cost" as "what will it do." Our aim has been not how much we can build, but how well we can build, positively assuring this personal attention to every detail of manufacture. Building for cheapness alone and disregarding excellence and quality is the most frequent and certain cause of loss of confidence and consequent business.[4]

Recognizing that a quality tractor would require quality materials, C. L. Best employed a high percentage of alloyed steel in the Tracklayer. The science of metallurgy was just beginning to find applications in the tractor industry. Best embraced these advances.

With each subsequent tractor designed and produced by the C.L. Best Gas Traction Co., considerable experience and knowledge was gained. Each model was field tested under everyday use. Careful notice was taken of the strengths and weaknesses and applied to future models. The tractor that earned its place as a truly transitional model was the Best Model A Sixty. The first generation of Model Sixty Tracklayers spanned serial number (S/N) 101A to S/N 1125A. Designed to be a 60 H.P. tractor, it retained a number of the strong features from the Model 75: the transmission used cut and machined steel gears running in oil, the tracks featured rocker joint links, the roller frames were of the oscillating type, alloyed metals were used as needed, and the motor was of a similar design to the Model 75. The Model 30 and Model 40 Tracklayer main clutch concept was used in the Model Sixty, while the Model 25 shared its use of antifriction bearings. The steering clutches from the Model 25 migrated to the Model Sixty with a proportional increase in size to match the increase in horsepower.

While the Model Sixty built on its predecessors, Best also incorporated many innovations into the tractor. Key to the success of the Model Sixty was the way the weight of the tractor was distributed. The balance of the tractor put the weight of the machine on the tracks. This allowed the tracks to stay flat upon the ground. That balance also enabled the Model Sixty to retain tractive effort by not lifting up in the front when under a heavy load. As a company catalog stated: "They stay down on the ground and pull." The weight distribution of the Model Sixty also permitted effortless steering.

Best applied advancements in heat treatment methods to the Model Sixty. These methods produced a better quality, longer-lasting tractor. With tractor S/N 966A, C. L. Best finally had the track design that later would become the industry standard. Even though his rocker joint link was serviceable, the development of a drop-forged, heat-treated track link proved to be the ultimate solution. The now heat-treated pin and bushing became fully round. The ends of the pin and bushing combination were pressed into the track links. This combination had enough clearance to allow the track to revolve around the front idler and sprocket with minimal friction. The next application of forging and heat treatment was to the track rollers. This virtually eliminated roller breakage and substantially increased undercarriage life. In a letter written many years later, shortly after the death of C. L., Ted Halton Sr. of Halton Tractor Company in Merced, California, shared a story as it had been told to him:

He [Fred Wass] had taken Mr. Best and his son and daughter and they started from Fish Camp early in the morning on a planned two week pack trip and they had stopped at Royal Arch Lake on their first night out. Fred said that at about two o'clock in the morning, Mr. Best came and got him out of bed and asked him to build a bright enough fire so that he could have enough light and warmth to make some drawings and while Fred kept the fire going Mr. Best drew a track shoe with a pin that turned inside of a bushing—one bushing being connected to one pair of links and the pin being connected to the next pair of links. Mr. Best told Fred that this would revolutionize the track laying tractor business and when morning came they broke camp and C. L. Best went back to San Leandro. I have read recently that a patent today practically has to be an

4. C.L. Best Gas Traction Co., Catalog E (1913): 12.

Best Sixty Tracklayer, number 101A. On June 24, 1919, this tractor was shipped to the Milt Talbert ranch in the Collinsville, California, area. It remained in active farm service until 1938. The tractor was then used to pull a grader for local road maintenance. This photo shows 101A pulling six wagons loaded with sacked wheat for export to the terminal at Rio Vista, California. Talbert's two children are on the tractor seat. Special Collections, University of California Library, Davis

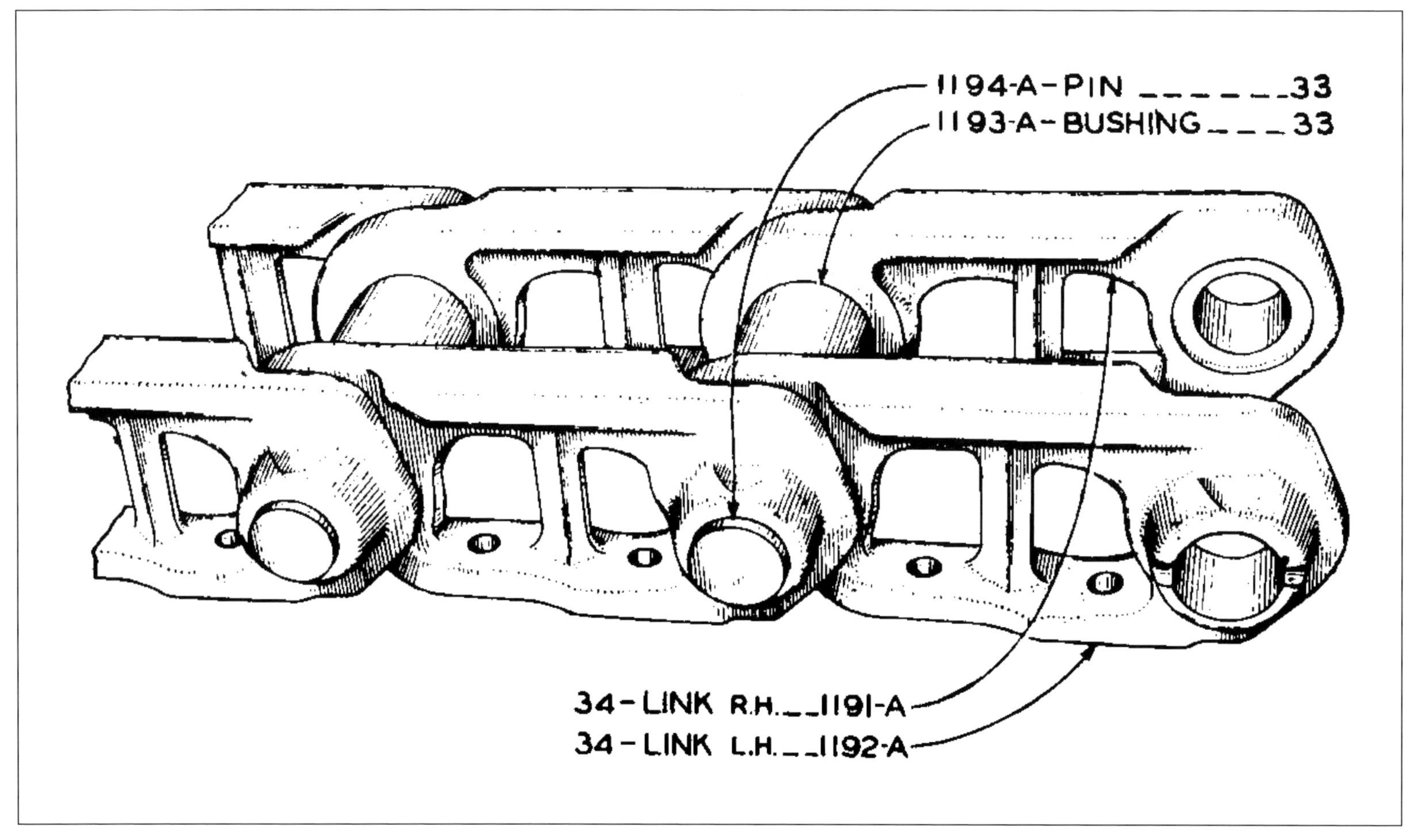

Diagram showing track link, pin and bushing, and associated parts numbers. Effective with Model Sixty S/N 966A, this patented C. L. Best design overcame the final problems associated with excessive track wear. Company parts book. Author's collection.

> *inspiration on the part of the inventor—it practically takes a stroke of genius. This comes close to that, doesn't it?*[5]

Antifriction bearings were used more liberally in the Model Sixty, totaling thirty-seven on the tractor. The incorporation of Timken bearings on the lower transmission shaft and the track sprockets allowed for finer adjustment and greater lateral stability. The Timken bearings could be adjusted to compensate for wear.

The second generation of Model Sixty tractors started at serial number S/N 1126A and ran through S/N 2200A. The major design change involved the tractor's final drives. The final drives were the last gear process providing power to the tracks. These drives were now inboard, enclosed (similar to the transmission gears), and ran in oil. Another change to the track drive was the separation of the sprockets from being part of the final drive. This allowed for replacement of individual parts when necessary instead of replacing the entire final drive unit. The track roller frames were also updated with the use of heavier structural steel.

With Model Sixty tractor S/N 2201A a final improvement was made to the tracks incorporating recoil springs. These

5. F. Hal Higgins Collection, Special Collections, University of California Library, Davis.

Best employee Ralph Easton is driving one of the first two Model Thirty tractors at the San Leandro factory test area. Note the smooth side channels on the radiator and the wire holding up the crank. Late 1920. Special Collections, University of California Library, Davis.

springs were mounted behind the front idler and would allow the tracks to contract when clogged with a solid object. This nearly eliminated breakage and excessive wear.

The final changes involved the engine. Formerly using the splash-type lubricating system, the Model Sixty now incorporated improved technology by using full-pressure lubrication. The tractor also saw its final boost in horsepower. This was accomplished by using off-center pivot rocker arms. Through its various design stages, the basic numbers of the Best Model A Sixty motor stayed constant: 6½" bore, 8½" stroke, and 650 RPM. C. L. Best accomplished what few other tractor manufacturers were able to do: The horsepower increases that his tractor received were through camshaft adjustments and rocker arm redesign, not by changing engine displacement and RPM. Best Tracklayer S/N 101A left the factory with 55 H.P.; Best Tracklayer S/N 2201A left the factory with 77 H.P.

The Best promotional materials stressed the quality and durability of the tractors offered for sale. As most prospective owners were familiar with the use of horses for power, it was natural to compare the new Best Tracklayer to a horse:

> *Did you ever see a race horse hooked to a plow? If you did, you saw him showing a burst of speed, pulling his heart out, and then after a short time, literally staggering with sheer exhaustion. Naturally he is a failure. He wasn't built for the plow and can't stand draft horse work. Neither can a race horse motor stand draft horse work. The Best motor was designed by Mr. C. L. Best. It is a valve-in-the-head heavy duty type—a draft horse motor built to do draft horse work. Design and construction are absolutely simple.*[6]

With the Model Sixty tractor providing power for the larger operations, the C.L. Best Tractor Co. soon produced a model more suited for mid-sized operations: the Best Model S Thirty. This tractor drew on all the strong points of the Model Sixty. The major innovation on the Model Thirty was the elimination of the engine frame and the bolting of the motor directly to the clutch housing of transmission case. This allowed the motor to rest on the equalizer bar giving the tractor proper three-point balance.

Few problems were left to be corrected when Best launched the Model Thirty tractor. The first tractors produced had problems with an underdesigned cooling system and breakage where the engine was bolted to the transmission case. The early Model Thirty tractors were also sold without recoil springs on the sides of the roller frames. These problems were corrected in later tractors.

While well-designed, competitive products were key to the success of the C.L. Best Tractor Co., a clear-thinking, loyal, and progressive management team was needed to provide guidance and leadership. Whether products or personnel, C. L. Best was known for making the right decision at the right time. The time for those decisions was at hand.

---

6. "Why Best Tractors Are Tracklayers," C.L. Best Gas Traction Co. (1920): 3.

The Right Men

With the Model Sixty ready for production, the Hawkins takeover bid resolved, and the Holt Manufacturing Company lawsuit settled, C. L. Best and his company were ready to launch into a new era of growth and prosperity. Having learned a harsh lesson from temporarily losing control of the company he founded, Best was conscientious in his choices for his inner circle of advisors. From finances to factory, the group of five that came together to form the backbone of the C.L. Best Gas Traction Co. stayed together for over thirty years.

Born on September 17, 1883, in Oil City, Pennsylvania, Harry H. Fair moved to California and was an investment broker for the San Francisco investment banking firm of Cyrus Peirce & Company. Given the proximity of San Francisco to San Leandro, Fair was able to follow the performance of the Best Company and became a shareholder himself. C. L. Best, seeking the necessary financial backing to continue to grow his company, acted on a recommendation by J. F. Carlston, president of the Central Bank of Oakland, to contact Cyrus Peirce & Company to plan for the long-term financial needs of the company.[1] Fair worked closely with the Best Company on financial matters.

In 1919, Harry H. Fair became a director for the C.L. Best Gas Traction Co. and a member of the executive committee. He devoted his considerable knowledge of the investment banking business to the advancement of the tractor company. By definition, an investment bank raises capital, trades securities, and manages corporate mergers and acquisitions.[2] By the middle of the 1920s, Harry H. Fair would take the C.L. Best Co. to an unprecedented position of power in the tractor manufacturing industry.

Byron Claude (B. C.) Heacock was born in 1889 near Emporia, Kansas. As a youth, he showed an aptitude for facts and figures. Even though his formal education ended at age fourteen, Heacock contin-

1. Oscar L. Starr interview, F. Hal Higgins Collection, Special Collections, University of California Library, Davis.

2. "Investment Banking," Wikipedia.com (http://en.wikipedia.org/wiki/Investment_banking).

ued his personal pursuit of knowledge through correspondence courses and extensive reading. After he had held various manual labor jobs, Heacock's strength with organization and accounting was recognized. In 1910, he was employed as a clerk in the First National Bank of Manhattan, Kansas. While still located in Manhattan, Heacock served as assistant purchasing agent for the Kansas State Agricultural College.

Following a move to Michigan, Heacock was employed as a purchasing agent for the University of Michigan from 1914 to 1916. During the next three years, his advancement in the accounting firm of Ernst & Ernst led to a position of assistant manager of the Detroit office.

The year 1919 saw B. C. Heacock, wife Nellie, and daughter Helen on the move. San Leandro, California, was the destination. As a new employee at the C.L. Best Gas Traction Co., he called upon past experiences in the accounting field and as a purchasing agent to provide valuable services to the company. As that company moved forward with its plans for innovative products and fiscal responsibility, the steady hand of B. C. Heacock was felt in an ever-widening circle.[3]

From Michigan to the Territory of Alaska to the Bay Area of California, the man who Raymond Charles (R. C.) Force was in 1919 was fortified by the experiences gained in all those places. R. C. was born in Croton, Michigan, in 1880. His father, George, set an example that would serve as a model for the younger Force in his adult years:

> *As a business man, he is regarded by the community as one of the keenest in the township; and one whose integrity is implicitly to be relied upon, and whose financial operations are conducted with the strictest justice to all concerned.*[4]

With formal training as a stenographer, the younger Force put his skills to use in the employ of the King Milling Company. By the spring of 1899, the prospect of adventure led him west to Washington and a position with the American Transportation & Trading Company. While employed by this company, Force spent time in the Fairbanks, Alaska, area. In 1904, Florence Heilig was the first teacher of the public school system in Fairbanks.[5] In 1905, she and Force were married. By 1912, the now five members of the Force family were in California. R. C. became associated with other family members in the California Corrugated Culvert Co. located at Berkley.[6]

When the United States entered into World War I in 1917, it became apparent to the U.S. government that tax revenues would not cover the expenses associated with winning the war. The Treasury used a series of bond issues to supply the funds needed. These were known as Liberty Loans.[7] R. C. Force worked with the Liberty Loan drives in the Oakland area. It was through these drives that he became acquainted with Harry H. Fair. Fair in turn suggested Force for a position in the C.L. Best Gas Traction Co.[8] In 1919, R. C. Force became an officer of that company and began the career that would ultimately allow him to assume a prominent role in giving track-type tractors to the world.

Native Californian Oscar Link Starr was born in San Francisco on September 29, 1885. By the time he was sixteen, he was an apprentice machinist with the Union Iron Works, a Bay Area manufacturer of steam engines and boats. After working for various concerns in the area, Starr was hired and spent nearly six years working for the Gorham Engineering Company in Alameda. At Gorham, he was made superintendent and oversaw the building of gas engines. The Gorham company built the first gas self-propelled fire engine with a turbine pump.[9] Starr recalled that engine:

---

3. "Former Caterpillar President Heacock Dies," *Peoria Journal Starr* (March 25, 1975).
4. *The City of Grand Rapids and Kent County, Mich. Up To Date* (Logansport, IN: A.W. Bowen & Co., 1900): 688–89.
5. "Mining in Fairbanks Area Was Different than in Yukon," *Fairbanks Daily News-Miner* (July 20, 1971): 6.
6. "In Memorium—Raymond C. Force by Allen L. Chickering," *California Historical Society Quarterly* (December 1952): 89.
7. "Liberty Loans," Answers.com (http://www.answers.com/topic/liberty-loans).
8. Oscar L. Starr interview, F. Hal Higgins Collection, Special Collections, University of California Library, Davis.
9. Grace H. Ziesing, editor, *From Rancho to Reservoir: History and Archaeology of the Los Vaqueros Watershed, California* (Rohnert Park, CA: Sonoma State University, 1999): 152–53.

It was a 6 cylinder engine. We took it to New York, put it through a 30 day and night pumping test and sold it to Seagrave.[10]

Having worked on engines for boats and motors for land vehicles, Starr then turned his ambition to the sky. In 1910, while partnering with Bill Gorham, he developed a two-cylinder radial airplane engine. As told by Starr:

It flew, but we got orders from Gorham's father to stop before we killed ourselves. The engine was sold to Stanley Hillar, father of the Helicopter name of today.[11]

After leaving Gorham, Oscar was hired by the Aurora Engine Company of Stockton, California, in April 1912. It was at this company, a subsidiary of the Holt Manufacturing Company, that his talent for production management became apparent. Starr immediately noticed that the shops did not have sufficient tools and that the workers were ill trained on how to operate the ones that were available. The company had hoped to produce two engines per day. In short order, Starr had gotten production up to three engines per day. By early 1913, Holt management was pleased with the progress and incorporated the Aurora Engine Company into the Holt Company. Oscar Starr, however, was not pleased with that turn of events and left the Stockton company. His next place of employment would be in Elmhurst with the C.L. Best Gas Traction Co. Starr related:

I had never met C. L. till then, though he started up here in 1910. I had PRODUCTION here. One or two of his Best Tracklayers were in production at the time. Best had so many financial problems at this stage that he wasn't paying much attention to building them. I started with one a month, soon stepped up to one a week; then two or three a week. We were paid once a month when Best had the money. Getting that payroll together once a month was a problem that called for selling a tractor or borrowing money.[12]

From 1913 to 1917, Starr was the Best factory manager. During his tenure, new tractor models were introduced, refinements in design and production were incorporated, and the manufacturing plant was moved from Elmhurst to San Leandro in 1916. But the company had its share of troubles. The protracted patent lawsuit with the Holt Manufacturing Company was still in the courts. It was costing the company a great deal of money and distracting management from the basic business of building tractors. The spring of 1917 saw the control of the company shift from C. L. Best to C. A. Hawkins. Hawkins had purchased 51 percent of the company from investors. Oscar Starr was not pleased with the change in management. With Hawkins determined to produce tracklayers in Dayton, Ohio, Starr was sent East to inspect the facilities and set up production.

At the same time, in East Peoria, Illinois, Pliny Holt of the Holt Manufacturing Company was having difficulty fulfilling the wartime orders for the Holt Caterpillars placed by the British and French governments. Knowing Starr's reputation for increasing production, Holt sought him out and asked for his help. Starr's response to Holt's request was, "I'm ambitious; make it enough money and I'll come out and help." Pliny Holt did, and Oscar Starr became a Holt employee.[13] Starr was placed in charge of tractor production and was credited with increasing output from two to fifteen tractors per day. He also oversaw the erection and installation of machinery in several new buildings.[14]

In California, circumstances were changing for Best and his company. Best took back control of his company from Hawkins in March 1918. The prototype Sixty Tracklayer was built. With being shut out of wartime tractor production, Best concentrated on the domestic market and was increasing sales and service networks. Things were on the upswing for the San Leandro com-

---

10. Oscar L. Starr interview, F. Hal Higgins Collection, Special Collections, University of California Library, Davis.

11. Ibid.

12. Ibid.

13. Ibid.

14. "Starr from East Speeds for West," *3 B's: Best's Busy Bunch 1*, no. 1 (1918): 1.

Best Sixty 101A at Walla Walla, Washington, on its introductory tour of the western United States. The demonstrations spanned April 23, 24, and 25, 1919. Special Collections, University of California Library, Davis.

pany, when, due to personal reasons, Oscar Starr made a trip from Illinois to California. Starr's mother, Rebecca, died in Hayward on June 5, 1918. While in the area, Oscar stopped to see his friend, Leo Best. Joe Sabatier, a Best salesman, remembered the meeting:

> *I recall when Starr came back to Best from Holt. It was in 1918 and I was to go to Cuba. I was translating for the Cuban dealer when Starr walked in. As he and Best renewed old acquaintances, kidding back and forth, C. L. says, "What about coming back?" Starr answered, "I might if you make it worthwhile." So the deal was worked out to give Starr the salary he asked. Don't know what he got, but I think it was the top salary of the Best concern.*[15]

Within six weeks of his conversation with C. L. Best, Oscar Starr ended his employment with the Holt Manufacturing Company and headed back to San Leandro, ready to renew his association with the Best Company. He resumed his position as factory manager. The delight at his return was well stated in the company newsletter:

---

15. Joe Sabatier interview (1957), F. Hal Higgins Collection, Special Collections, University of California Library, Davis.

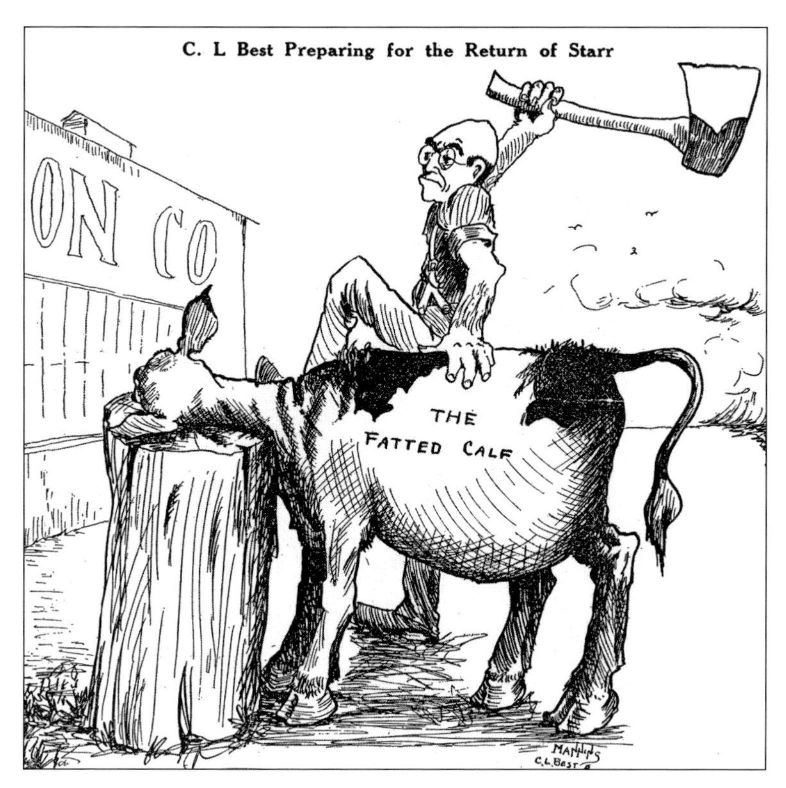

This cartoon appeared on the front page of the Best's Busy Bunch, *the company newsletter published by the C.L. Best Gas Traction Co., and demonstrates the importance of Oscar Starr returning to San Leandro and the Best company.* Author's collection.

> *The true test of the value of a man is the degree in which he is missed when he is gone and the warmth of his welcome back. While all credit is due those who kept the wheel in motion during his absence, yet we have all missed the man who is held in highest regard by the present official and employees throughout the entire plant.... Welcome to California. May your incumbency with the Company be pleasant, profitable, and redound to the mutual benefit of all concerned.*[16]

There were many lesser-known employees who loyally stood by the C.L. Best Gas Traction Co. from its modest beginning in 1910 in Elmhurst, through the difficult times in the late 1910s, and finally to 1919 when people, product, and production came together to take the company to a new level of prosperity.

Byron Williford started with the Best Manufacturing Company in 1905. At nineteen years of age, he realized he wanted more for himself than just farming with his father and brothers on their property in Bradley Township in Monterey County. When applying for a job, he specifically wanted to see the boss, but Williford had trouble believing the man he was taken to see really held that position. Earlier that day, C. L. Best had been involved in an accident at the plant. The *Oakland Tribune* reported on the incident:

> *A painful accident was suffered lately by Leo Best, the son of D. Best, proprietor of the Best works. He was engaged in testing a new traction engine when a tank of gasoline was exploded and the flames blown into his face.... It was found that the young man's face was severely burned, but the eyes were not permanently injured. The injury was exceedingly painful, and it will be several weeks before the wounds heal.*[17]

Finally convinced that he was truly seeing the boss, and asked what he could do, Williford said he "could work." He was given a job in the Best shipping department and proved that he wasn't afraid of hard work. That work ethic advanced him to the parts department. While the Best Manufacturing Company remained his day job, Williford became an electrical engineer by attending night school and completing correspondence courses.[18] C. L. Best gave Williford credit for the concept of the electrical combined harvester built by the Best Gas Traction Co. while it was in Elmhurst. Best also spoke of the fierce loyalty that Byron Williford had for the early and later Best Companies, even to the point of fisticuffs with ranchers who besmirched either the product or the producer.[19]

16. "Starr from East Speeds for West," *3 B's: Best's Busy Bunch 1*, no. 1 (1918): 1.

17. "Burned by an Explosion," *Oakland Tribune* (June 1, 1905): 12.

18. "I Left the Farm to Build Tractors," *Caterpillar News & Views* (September/October 1954): 23.

19. F. Hal Higgins Collection, Special Collections, University of California Library, Davis.

Tracklayer 101A shown in a field demonstration in Colorado. Seated on the tractor is Byron Williford. Special Collections, University of California Library, Davis.

When C. L. Best was ready to incorporate emerging technology and advanced design into a tractor that would leave the other manufacturers behind, Byron Williford was on the forefront of both development and testing. Before it went into production, the Best Sixty Tracklayer spent months in field trials and farm tests, with Williford staying with the tractor to monitor its performance and make corrections and repairs. When the tractor was ready for production, Williford was named superintendent of the machine shop.[20] Byron Williford helped set the tracks in motion that would establish the standard for track-laying tractors of the future.

Born in Chicago, Illinois, in 1877, Walter Grothe spent his early adult years in Ohio. Grothe's father was employed by Rollin White, owner of the White Truck factory in Cleveland—the same Rollin White who purchased the Best 8-16 "Pony" Tracklayer displayed at the 1915 Pan Pacific Exposition. As early as 1903, Walter Grothe was in California racing steam automobiles built by Rollin White.[21] It was this association with White that introduced Grothe to fellow racing enthusiast C. A. Hawkins. When Hawkins bought controlling interest in the C.L. Best Gas Traction Co. in 1917, Grothe made the move from Ohio to California. But while Hawkins only stayed with the company for one year, Walter Grothe found a strong position with the Best concern.

To be of any value to a farmer, logger, freighter, or other user, a tractor had to withstand the demands placed on it by the operation and still continue to function. C. L. Best believed that building a more expensive, stronger tractor that kept working was by

20. "I Left the Farm to Build Tractors," *Caterpillar News & Views* (September/October 1954): 23.

21. "Automobiles in Race," *Oakland Tribune* (August 6, 1903): 4.

far more economical to the owner than a less expensive tractor that failed to withstand the demands placed on it. One key to the superiority of the Best tractors was the pioneering use of metallurgy in tractor production. When the Sixty Tracklayer was in the development stage, it was realized that to achieve the mandate issued by Best concerning quality and longevity, ordinary steel would not be good enough. Walter Grothe was the point man for the Best Company when it came to adopting procedures for heat-treating metals.

Realizing how much there was to learn about these emerging metal treatments, Grothe spent many hours with Robert R. Abbott, metallurgical engineer for the Peerless Motor Car Company in Cleveland, Ohio. Abbott was credited with developing the early metallurgy in the auto industry. Grothe took this knowledge and continued to expand and refine the metals and procedures used on Best tractors. Carburized steel was put into use, as was the forging of steel to provide strength. Grothe recalled:

> *C. L. Best directed Engineering and Development himself. I was merely out there trying to make the transmission stand up. Shafts and sprockets needed more guts. There was always a weak spot some place that had to be strengthened to keep up with the demands on the tractor for tougher and harder work. We began to develop forgings to replace cast gears, track shoes and links. Spool and pins we began carburizing. Forging from bar stock came in then. Only ordinary steel was used up to this point. A lot of quality control went into the forging plant and steel mills for checking quality at the source. About 28 to 30 inspections were made at the source after we got the cooperation from the steel companies to meet the standards demanded. We got so we never let them ship us steel until [it was] checked at the source. We had some great battles at some plants.*[22]

By continually searching for improved metal treatment methods and ensuring the use of quality, standardized materials, Walter Grothe was a key player in making the Sixty Tracklayer live up to its name: Best.

Born on May 24, 1882, in Izaourt, France, Joseph John Sabatier left the family farm and France in 1899 and used his small savings to buy a ticket on a cattle boat to Norfolk, Virginia. When that boat left French waters, little did he realize that his tenacity and his acquired ability to speak three languages would send him around the world, selling tractors for a company that in the future would come to dominate that world. After docking in Norfolk and helping unload the boat, Sabatier took the train to the West Coast. During the next fifteen years, he learned English, married Grace Garat, became a naturalized citizen in 1912, and worked at a number of jobs. It was while he was employed in the Oakland area that he first saw a Best tractor and decided that he wanted to be associated with those tractors and that company. Sabatier told of his struggle to be hired at the C.L. Best Gas Traction Co.:

> *[I] called on Best for a job every day I had off from my job. Best was then in Elmhurst. Answer at Best's was always NOTHING. I moved to San Leandro and kept on going over to Best's. Finally, one day a fellow answered, "I got a job for you but you won't do it. Come back on Monday and bring your overalls. Report at Service." There I was handed a steel brush, putty knife and kerosene. In the yard were the old steam engines that came into the plant for overhauling. They had 18" rims with muck all over them. In three weeks I had that old engine clean as a whistle and reported to the office, figuring I would now get to do something better. But just then another one came in and I was put on that. So for three and a half months I did that kind of mean work at $1.25 a day.*[23]

That hard work and refusal to quit finally earned Sabatier a job in one of the Best shops at $3.50 per day. During the next four years, he did service work, demonstrated tractors, and also

22. Walter Grothe interview (March 7, 1952), F. Hal Higgins Collection, Special Collections, University of California Library, Davis.

23. Joe Sabatier interview (1957), F. Hal Higgins Collection, Special Collections, University of California Library, Davis.

(Above) *Automatic rotary hearth furnaces used in heat-treating track parts.* Company catalog. Author's collection. (Left) *C. L. Best employed various aspects of the science of metallurgy to improve the quality of his products. This row of heat treatment furnaces is located at the Best heat-treating plant.* Company catalog. Author's collection.

(Above) *Signage on the building reads "For Field and Orchard-Plowing and Heavy Hauling. C.L. Best Tractor Co." Note Best Sixty tractor on the loading platform. 1921.* San Leandro Public Library Historical Photograph Collection #01101. (Left) *Materials in the process of heat-treating.* Company catalog. Author's collection.

did anything else that the company needed. That anything else even encompassed troubleshooting the owner's father's car. Daniel Best had passed his seventy-fifth year and was still going on road trips in his Packard automobile. When he broke down coming back to San Leandro from a trip to Stockton, Joe Sabatier was dispatched to the scene. While it took him only fifteen minutes to get the automobile operating again, Joe made a lasting impression on Dan Best with his service.[24]

By 1918, Sabatier was out of the shops and into export work. His language skills were indispensable to the company. Many times he was called upon to translate for a visiting dignitary. While around the offices, Joe was witness to two notable events. He was there when the notorious fist fight between C. A. Hawkins and H. C. Montgomery took place. He was also in the office translating for the Best Company's Cuban dealer when Oscar Starr and C. L. Best made the deal in 1918 for Starr to come back to work for the company.

The Best company sent Sabatier to Cuba in late October 1919. With the Best Sixty in production, the company was hoping to expand its foreign markets, and Cuba seemed a likely choice. Cuba had large sugar cane plantations and needed a more efficient method of transporting the cane to the mills. Joe Sabatier needed a way to convince the plantation owners that a tracklaying tractor was exactly what the conditions demanded. One owner was willing to take the risk on the $6,000 tractor, providing Sabatier would remain in Cuba and demonstrate the tractor's abilities. When the Sixty tractor and wagons arrived from California, Joe upheld his part of the bargain and remained for five and a half months. When the rainy winter weather stopped the oxen from hauling cane to the processing mills, the Best Sixty kept working. Sabatier and his tractor so impressed the other plantation owners and their managers that the Best Company made over thirty sales in Cuba the following year.[25]

Joseph Sabatier continued to travel the world in the employ of the C.L. Best Gas Traction Co.; the Best Sixty continued to make impressive market gains, and C. L. Best continued his advance to dominate the track-type tractor field. But national and international economics in the early 1920s would put the decisions made by Best to the test. Could the C.L. Best Gas Traction Co. survive when the rate of business failures tripled? Would the San Leandro business suffer the 75 percent decline in profits seen by many of those companies that did survive?[26]

24. Joe Sabatier interview (1952), F. Hal Higgins Collection, Special Collections, University of California Library, Davis.

25. Ibid.

26. "Depression of 1920–1921," Wikipedia.com (http://en:wikipedia.org/wike/Depression_of_1920).

Post–World War I Years

The World War I years were profitable for those engaged in agriculture. With the war raging, agricultural production in Europe declined. This allowed for a substantial increase in exports from the United States, resulting in more land being farmed and a rise in farm product prices and incomes. In the track-type tractor market, the Holt Manufacturing Company first benefited from contracts with European governments, then from the U.S. government, to provide its Caterpillars to the war effort. While Holt expanded its export market, domestic sales sagged. The war years affected the C.L. Best Gas Traction Co. in an opposite manner. Shut out of war contracts, Best concentrated on the domestic market. He continually refined his tractor design, culminating with the production of the Sixty in mid-1919. With a solid tractor to offer buyers, and his management, production, and sales teams in place, the future looked bright for the company.

The year 1918 brought changes to the world. After the Armistice ending the war was signed that year, there was an adjustment period from war time to peace time. Manufacturing companies that relied heavily on government war contracts were severely affected when those contracts were abruptly canceled. The U.S. economy struggled to find balance through 1919. By 1920, unemployment reached 11.7 percent and the Dow Jones Industrial Average bottomed in August 1921 with a decline of 47 percent from its peak in November 1919. The Federal Reserve System's plan of action did little to help the situation. In December 1919, it raised the interest rate from 4 percent to 4.75 percent

Best Tracklayer display at the Washington State Fair. Tractors from left to right: Model Twenty-Five, Model Sixty 101A, and Model Seventy-Five. Note the radiator side channel on the Model Sixty has the words painted, not cast as was the case in later production models. April 1919. Special Collections, University of California Library, Davis.

and by June 1920, the rate stood at 7 percent. That high rate of interest dramatically reduced the amount of lending done to other banks and to consumers and businesses.[1]

The future looked bleak for those businesses without a clear vision of the future and the drive to achieve that future. C. L. Best was not lacking in either vision or drive. In his letter to the stockholders in the 1920 annual report, Best reported:

> *The closing months of last year [1920] witnessed a substantial decrease from the volume of sales which would have resulted had general conditions remained normal. The field of sale for Best Tractors has been extended both in this country and abroad and, by reason of such increased outlet, it has been possible to keep our plant running continuously while many industries engaged in similar manufacture have been obliged to either greatly restrict operations or entirely close down pending improved market conditions... and by increased effort and careful planning we look forward with confidence to the future.*[2]

The Model Sixty was the only tractor in production until the new model Thirty had completed it testing phase. With both the Thirty and its big brother, the Sixty, now in production, the San Leandro plant would need to be expanded and updated. As before, C. L. Best embraced manufacturing innovations that would allow his tractors to be produced more efficiently. With the market contracting, but with an unwavering faith in his product and ability to produce it, Best proceeded with the expansion

1. "Depression of 1920–1921," Wikipedia.com (http://en.wikipedia.org/wiki/Depression_of_1920).

2. "To the Stockholders," Annual Report C.L. Best Tractor Co. (April 7, 1921): 2.

Two Model Sixty tractors with their engines crated ready for shipment from the San Leandro factory. The lifting slings used to load the tractors onto the flat car are still attached to the tractors. Canopies are packed in the wooden crate between the tractors. 1923. San Leandro Public Library Historical Photograph Collection #01536.

Factory crane over Davis Street in San Leandro after it was repainted to reflect the company name change to C.L. Best Tractor Co. 1920. San Leandro Public Library Historical Photograph Collection #00851.

plans. Cyrus Peirce & Company acted as underwriter for the offering of $1,250,000 of preferred stock.[3]

An additional change to the company occurred in 1920. On October 15, the C.L. Best Gas Traction Co. changed its name to the C.L. Best Tractor Co. *Gas traction* had become an antiquated term originally used to differentiate between steam and gas tractors. The word *traction* had been used to designate a machine that could move under its own power as opposed to a stationary unit. For the tractor buyer of 1920, the name change was a representation of the focus on the future by the C.L. Best Tractor Co.

The year 1921 saw little relief from the economic down turn. The annual report published in March 1922 spoke of the condition of the company. While calling the manufacturing businesses in general "very much demoralized," Best touted the fact that his business continued "uninterruptedly the payment of dividends on our preferred stock when suspension of dividends was the order of the day." The business continued its fiscal policy of "extreme conservatism." But "in spite of adverse conditions, sales for the year 1921 exceeded 60 per cent of sales during 1920. While the net gain for the year was small, it is distinguished by the fact that heavy losses have been the experience of most manufacturers for the year 1921."[4]

Magazine and newspaper ads from 1921 show the approach used by the company to promote the Thirty and the Sixty:

> *Your investment in machinery this year should be with an eye to the future. The new equipment you acquire should be so inherently long-lived that you don't need to think of replacement for years to come.... Buy a tractor about which there is not the slightest doubt as to its long-wearing qualities.... Best is the safe tractor for you to buy this year—or any year, and this is the year in which a good tractor can help reduce the cost of crop production.*[5]

3. "Best to Enlarge Plant," *Oakland Tribune* (April 28, 1920): 17.

4. "To the Stockholders," Annual Report C.L. Best Tractor Co. (March 1, 1922): 2.

5. C.L. Best Tractor Co. advertisement, *Fresno Morning Republican* (September 25, 1921): 10A.

The C.L. Best Tractor Co. was represented by its local dealerships. Number one on a list of qualifications for a dealer was "good character and a sustained reputation for square dealing." The company felt that to be most beneficial, the manufacturer and dealer must have "unbounded mutual confidence in and respect for one another."[6] The dealerships played important roles for the company. They served as a buffer between the manufacturer and the tractor purchaser. The San Leandro headquarters could concentrate on production and new product development and leave the selling to the dealer. The dealerships were required to maintain showrooms to showcase sample tractor models. Also required was a significant investment to have ample parts and service men available when needed. This supported Best's strategy of keeping the tractor running and making money for the owner.

The difficult times of the early 1920s were felt from the factory to the dealerships. As a show of support to its dealers, the Best Company offered help:

> *Business became so badly depressed in 1921 that the Best dealers did not feel able to take used equipment in trade and as a result the dealers stopped trading, which really meant that they practically stopped selling. Tractors were piling up in the backyard at the factory in very uncomfortable numbers. In order to set the wheels again in motion, the Best Company offered to assist its dealers by participating equally with them in any amount that the dealer might allow for a used tractor in connection with the sale of a new tractor. The effect of this offer was instantaneous and from that moment, sales began to pick up.*[7]

Combined production figures for the Thirty and Sixty tractors in 1922 nearly doubled the amounts produced in 1921. The tractors taken in trade also provided the dealers with a lower-priced alternative for buyers who saw the wisdom of tractor ownership but could not afford a new model. Buyer loyalty to the Best brand was sought and encouraged. In the stockholder report for 1922, President C. L. Best noted that "through a carefully selected dealer organization in this country and abroad, it is gratifying and significant to observe that repeat orders are the rule."[8]

While some of those repeat, as well as new, orders were for agricultural use, Best dealers were expanding the scope of their sales. From the time of Daniel Best's steam-traction engines, the Best name stood for quality and dependability to the lumbermen. The Tracklayer continued this tradition. Mine and quarry owners saw the tractors as worthwhile money-making investments. Building a bridge? Clearing the land for a new subdivision? Building a new stadium or golf course? Moving a house? Completing sewer projects? Operators put the tractors to a multitude of different uses, but Henry Ford and his Model T demanded a use that would become synonymous with crawler tractors: road building.

When Ford presented his automobile to the public, roads were little more than tracks and trails. Over the years, as autos became more affordable, the public demanded better roads to travel. Federal and local governments were the agencies that provided the funding for those roads, and Best tractors provided a good source of power to build those roads. By 1920, having resolved the undercarriage problems that had been inherent to track-type tractors, C. L. Best offered a machine that would withstand the rigors of building roads and highways and just about anything else asked of it. To get the information out to prospective buyers, the company placed its advertising account with K.L. Hamman Advertising of Oakland.[9] Focusing on the success of the Best tractors with regards to road building and maintenance, the firm inserted numerous ads in *American City Magazine* and similar publications. With multipurpose machines, a state-of-the-art manufacturing facility, and an expanding network of dealerships, the C.L. Best Tractor Co. was well situated to take bold steps when the U.S. economy began to rebound in 1923.

6. *Across the Table* (San Leandro, CA: Caterpillar Tractor Co., October 1926): 6.

7. Ibid., 78.

8. "To the Stockholders," Annual Report C.L. Best Tractor Co. (February 26, 1923): 2.

9. "Hamman Takes New Accounts," *Oakland Tribune* (October 15, 1922): 1.

Building the road bed of Federal Highway 12, near Howard Lake, Minnesota, with a Best Sixty and elevating grader. Circa 1925. Author's collection.

(Top left) *This rear seat tractor with the operating controls on the right-hand side for agricultural use (plowing, harvesting, etc.) was the standard configuration from tractor serial number 101A to 1125A.* Company catalog. Author's collection. (Bottom left) *The top seat adaptation of the Best Sixty was a result of increased use by logging operations. The top seat location provided for improved operator safety and a smoother ride. Tractor serial numbers were interspersed within the 101A to 1125A range.* Company catalog. Author's collection. (Bottom right) *With increasing highway traffic and tractor usage for road maintenance, the Contractor's Special was offered with the seat and controls shifted to the left-hand side. Tractor serial numbers were interspersed within the 101A to 1125A range. A Swamp Special tractor was considered but it is likely that it never went into production given that the parts needed for this version do not appear in parts books.* Company catalog. Author's collection.

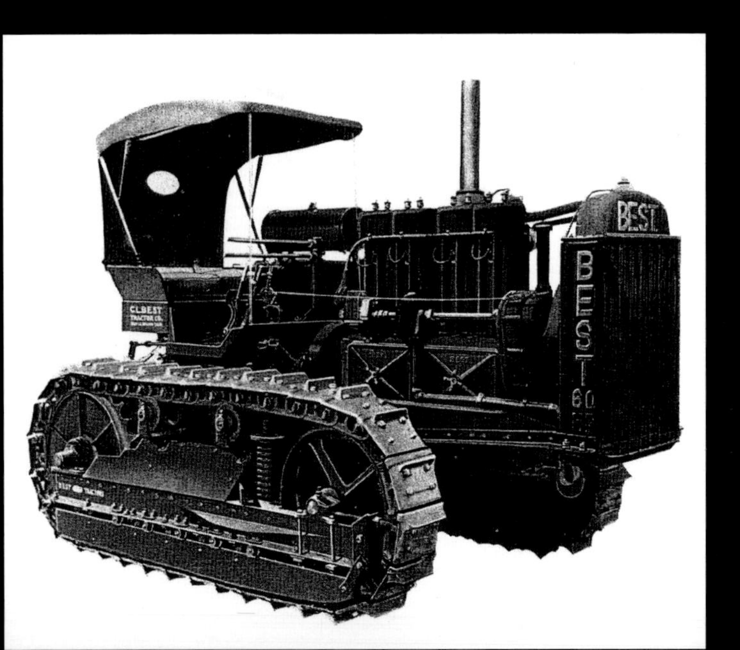

The Beginning of an End

The year 1923 saw a marked improvement in the health of the U.S. economy. Unemployment rates dropped throughout the year, ending at 3.3 percent.[1] With more people working for good wages, there were demands for modern goods and services. The expanding availability of electricity allowed many labor-saving appliances into households. Refrigerators, vacuum cleaners, electric irons, washing machines, and radios were just a few of the modern goods. The boost in the economy allowed for more ownership of single-family homes and automobiles. The use of trucks to move goods to market expanded. California saw an increase of over $2 million in the motor vehicle fees collected in 1923 compared to 1922. After operating expenses were deducted, the remaining funds were used for road upkeep.[2] The Best tractors provided much of the power for that upkeep.

As the versatility and reliability of the Thirty, Sixty, and Cruiser tractors became widely known, demand for the tractors showed a steady increase. Profits for the company did likewise. Money continued to be spent on the manufacturing facility. In June 1923, the C.L. Best Tractor Co. invested $60,000 in overhead fire protection. The system had nearly 20 miles of pipes with 20,000 sprinkler heads and covered the entire plant. The system was supplied with water from two 50,000-gallon gravity flow tanks.[3]

---

1. John Dunlop and Walter Galenson, editors, *Labor in the Twentieth Century* (New York: Academic Press, 1978): 98.
2. "Over Two Million Dollar Increase in Motor Vehicle Receipts in 1923," *Woodland Daily Democrat* (February 27, 1924): 8.
3. "Fire Guard for Plant to Cost $60,000," *Oakland Tribune* (June 17, 1923): 1.

The annual report for 1923 noted that a common stock dividend of 100 percent was declared, payable in stock, which brought the total of common stock to $1,000,000. The sales for the year topped $4,000,000 while profits were $703,921. In the report, certified by Price, Waterhouse & Company and made public by Peirce, Fair & Company, Best management stated that it was the best year in the company's history. Indications for the coming year, 1924, were favorable.

The 1923 annual report dealt with the financial progress of his business, but C. L. Best suffered an immense personal loss that year. On August 22, at the age of eighty-five, Daniel Isaacs Best died after a short illness. With his passing, C. L. lost more than a father. Daniel was C. L.'s early mentor and hunting partner. He was a wise counsel through C. L.'s difficult times building his business. Daniel and C. L. spoke the same language: inventing and manufacturing. Sitting by the fireplace in the big Victorian house on Clarke Street, Daniel with his pipe and C. L. with his cigarette, hours were spent discussing problems, solutions, and what the future would hold. Years of hard work and difficult decisions were paying off: The Best Sixty tractor was in its fifth year of production, and the Thirty was making a strong showing. Before he died, Daniel saw what the future would hold for C. L. and the C.L. Best Tractor Co.: to simply be the best.

The growing prosperity of 1923 continued into 1924. In May, the company called for redemption of its preferred stock issue of $1,250,000 and offered an exchange for like par value of its capital stock. In a letter from Peirce, Fair & Company to stockholders, the investment bankers stated that

> *we believe that the best interests of our clients will be served by taking advantage of the company's offer to exchange rather than accepting $100 a share for the preferred stock. The company enjoys exceptional management, and from our close connection with it we feel it promises a remarkably prosperous future.*[4]

4. "Preferred Stock of Best Tractor Callable May 1," *Oakland Tribune* (April 24, 1924): 26.

Daniel Best. This photograph is signed "To Leo from Dad." **Courtesy of Dan G. Best II.**

With the finances of the C.L. Best Tractor Co. being overseen by Peirce, Fair & Company, C. L. Best and his sales staff worked hard to show the versatility of the Best Tracklayers. Even though the machines were built in sunny San Leandro, they soon proved their worth in the winter snows.

> TRACTOR BUILT IN EASTBAY TOPS
> PIKE'S PEAK CLIMB
>
> *A Best Tracklayer Tractor... has just cleared the snow and ice from the automobile highway which runs to the top of Pike's Peak. The demonstration was requested by the Pike's Peak Auto Toll Road company who thought it would be of great financial advantage if the road could be kept open later in the fall and cleared earlier in the spring. The*

(Top) *Best Sixty Snow Special with a LaPlant Choate snow plow sold by the Wm. H. Ziegler Co. of Minneapolis, Minnesota. Photo taken in front of the Ziegler building. To demonstrate to the State of Minnesota the benefits of open roads in the winter, Ziegler plowed Federal Highway 52 from St. Paul nearly eighty miles to the City of Rochester.* Photo courtesy of Ziegler, Inc. (Lower left) *Head-on view of ramming snow drifts on Federal Highway 52 in Minnesota.* Photo courtesy of Ziegler, Inc. (Lower right) *View of cut made by the Best Sixty tractor on Federal Highway 52. Note the extension of the snow plow wings.* Photo courtesy of Ziegler, Inc.

Best Thirty tractors at the C.L. Best Tractor Co. factory. 1921. Photo courtesy of Martin Breitmeyer.

> *Best proved equal to the task and the Toll Road company bought the tractor.*[5]

The Best Snow Special tractor faced great difficulties on Pike's Peak. The snowline began at 11,400 feet. Freezing and thawing hardened the snow drifts that the tractor faced for ten miles on its trip to the top. The thinner air of the higher altitude decreased the horsepower of the engine. The highway, with its 10.5 percent grade, was completely covered by snow. Great caution and skill from the operator was needed to keep the tractor from plummeting down the mountainside.

When asked by reporters to comment on the Pike's Peak climb, Best spoke with confidence about the roles his tractors could play in clearing snow-blocked roads:

> *"When we were asked if our tractor could accomplish the task, we had no hesitancy in promising that the job would be done," said C. L. Best, president of the C.L. Best Tractor Co. "We have felt for a long time that closing the five passes over the Continental Divide for 9 months each year, blockades of snow on the highways of the Sierras, closing of the Pacific Coast highway from Mexico to Canada because*

5. "Tractor Built in Eastbay Tops Pike's Peak Climb," *Oakland Tribune* (June 1, 1924): 13.

Close-up of Best Thirty tractors at the factory. Person unidentified. 1921. Photo courtesy of Martin Breitmeyer.

of snow, were all unnecessary. Even Yosemite could be open to the motorist either winter or summer."[6]

While snowplowing in the Rockies was dramatic, the Best tractors moved mountains of snow from roads all over the nation. Many times teamed with LaPlant-Choate plows, the tractors adapted well to this wintertime use. With its enclosed wooden cab, powerful headlights, change in fuel tank placement, and a hood to help retain motor heat, the Best Snow Special tractor logged countless hours in snow removal service to state highway departments and other local units of government. For many people east of the Rocky Mountains, their first glimpse of a track-type tractor was doing road work: either in front of a pull grader maintaining roads and streets or behind a mounted V plow with wings clearing the roads after a snowstorm. Instead of being restricted by the winter snows, commerce would now be able to carry on year round. The sound of a Best Tracklayer clearing the roads on a crisp winter night announced to the farmers and ranchers that they were no longer isolated; the trip to town would be much easier thanks to the power of the Tracklayer.

As owners and operators found more and varied uses for the Tracklayers, the Best dealer network was also expanded. In 1919, the company had fifteen dealers, mainly on the West Coast. By 1924, that number had increased to fifty domestic dealerships

6. Ibid.

with seven export outlets.[7] Quality dealerships supported the Best tractors with parts and service when necessary and acted as intermediaries for the customer and the factory. This allowed the customers to keep producing and profiting and the factory to keep refining and manufacturing the products those same customers demanded. Production figures for the Best concern showed a significant increase of 26 percent in the number of Sixty tractors produced in 1924 compared to 1923. The numbers for the Thirty were even more astonishing: a 91 percent increase in the production of the smaller tractor.

One way the Best Company kept refining its products was to conduct a yearly review of both current and newly offered equipment. From an article in the *Oakland Tribune*:

> *Every year the Best Company reviews the equipment used on its tractors and subjects each unit to rigid competitive tests, even where the unit has been standard equipment the year before. This plan assures newly developed products a chance in the competition, and makes old established ones prove their "mettle" anew every year. It keeps the engineering department on the alert. The success of the plan may be judged by the fact that in the last 60 days the Best company has increased its production schedule 40 per cent.*[8]

As the C.L. Best Tractor Co. continued to prosper, others were noting its success. Best's major competitor, the Holt Manufacturing Company, clearly saw the smaller company as a threat. In a Peoria sales report from 1924:

> *I thoroughly believe that 1925 [will be] a most critical year. If the Best company makes as much progress in the next 12 months as it has in the past year or two, our volume... will be menaced.*[9]

Holt had always been the larger of the two firms. By the mid-1920s, its combined plant facilities of Stockton and Peoria were six times the size of the Best production facility at San Leandro. While Holt sales were still twice as much, the gains by the C.L. Best Tractor Co. would indeed turn 1925 into a critical year for Holt.

The financial world also noted the ascendancy of the Best company. Numerous times the company found itself in the headlines: "Best Tractor Continues to Rise," "Best Tractor Co. Reports Banner Year," "Best Tractor Jumps," and "Tractor Company Has Good Earnings." When the annual report for 1924 was issued, the San Francisco financial district was pleased. Sales for 1924 reached $5,344,000 as compared to 1923 with approximately $4,000,000. Net earnings, before federal taxes, were up nearly 30 percent from 1923 to $901,830 in 1924. As far as assets to liabilities, the 1924 report set those at seven to one.[10] Late in 1924, the *Oakland Tribune* noted that Best Tractor was "in new high ground. The continual rise in Best Tractor... is a reflection of the popular favor with which this local industrial is regarded and of the promising outlook for larger earnings."[11] Also noted by the *Tribune* was the fact that the C.L. Best Tractor Co. had awarded three extra dividends in 1924. The final extra dividend was 25 cents per share. This was offered in addition to the regular quarterly dividend of $1.25 per share.[12]

With a solid year of growth behind them, C. L. Best and his executives were looking forward to 1925, but that forward look didn't stop with 1925. Where would the company be in one year, five years, ten years? With its premier products, expanding production, solid financial backing, and progressive corporate leadership team, the C.L. Best Tractor Co. kept moving forward in its drive to be the dominant track-type tractor manufacturer. But would that drive also lead the Best Company to the East?

---

7. *Fifty Years on Tracks* (Peoria, IL: The Caterpillar Tractor Co., 1954): 28.

8. "Tractor Firm Test Product Most Rigidly," *Oakland Tribune* (December 7, 1924): 9.

9. *Fifty Years on Tracks* (Peoria, IL: The Caterpillar Tractor Co., 1954): 28.

10. "Tractor Company Has Good Earnings," *San Francisco Chronicle* (February 20, 1925): 23.

11. "Best Tractor Continues to Rise," *Oakland Tribune* (December 23, 1924): 28.

12. "Dividends," *Oakland Tribune* (December 11, 1924): 29.

And the Winner Is . . .

In 1919, C. L. Best had put together a management team that embodied his vision for the future of his company: a company that would provide quality products and service at a fair price to a market that would extend from coast to coast and also include the world.

When the financial trials of the early 1920s ended, Best was well on his way to accomplishing most of what he had planned for his company. The Sixty and Thirty tractors were in production and considered to be groundbreaking in design and performance. While not inexpensive to build or buy, the durability and cost effectiveness of the Best tractor was proven again and again by the myriad of uses found for it. The network of domestic dealerships continued to be expanded. To provide support to the eastern dealerships, Best established a parts warehouse in St. Louis, Missouri. The export market was also increased.

While Oscar Starr kept the production high at the San Leandro plant, there was one obstacle that even Mr. Starr's expertise could not surmount. With only one manufacturing plant located on the Pacific Coast, transportation of products and materials became a pivotal issue. The steel mills that provided the basic essentials for tractor production were located over two thousand miles to the east. Completed tractors sold by dealerships in the central to eastern sections of the country faced a similar two-thousand-plus-mile journey. The cost for the journey east could be considerable for a tractor the size of a Sixty. Even with this added transportation expense, the Best Tracklayers were still increasing their market share against their nearest competitor.

The Holt Manufacturing Company was the major competitor of the Best concern, but it was not the only competitor in the track-type tractor field. Other companies were also producing track-type tractors. Three examples were Cletrac in Ohio, Bates Steel Mule from Illinois, and Monarch built in Wisconsin. To be able to effectively answer these challenges, C. L. Best had to move production to the east. Although the attempt to set up a plant in Dayton, Ohio, during the 1917 Hawkins period was a fiasco,

Rear seat Best Sixty tractors being prepared for rail shipment at the San Leandro factory. 1921. Author's collection.

C. L. Best, along with H. H. Fair, R. C. Force, O. L. Starr, and B. C. Heacock, was acutely aware of the advantages of a manufacturing plant east of the Rockies. St. Louis, while functional as a site for a parts depot, was too distant from the source of the customized metals used in the tracklayers to be a viable site for a production facility. Best and his inner circle held cautious, informal talks about how to acquire more easterly facilities. By 1924, the decision had been made: the state would be Illinois, the city would be Peoria, and the facilities would be those owned by the Holt Manufacturing Company.[1]

The Holt Company had experienced a difficult transition from the production of tractors for World War I to the production of tractors for the domestic market. By 1923, Murray Baker, a Holt executive of the Peoria facility, noted that competition from the C.L. Best Tractor Co. was the most difficult problem he faced. The Holt sales department felt the need to provide its salesmen with a twenty-five-page document answering, point by point, each of the nineteen arguments used by the Best sales staff to promote Best products. When the findings of a study commissioned by the Holt Company reached President Thomas Baxter's desk in 1923, it contained specific recommendations to ensure a dynamic future for the company. The most notable problem according to the report was the shortfall in sales. This resulted in the factories operating at just 60 percent of capacity. The report further recommended changes to the number and models of tractors and harvesters that should be produced and an increased use of parts commonality. Peoria was seen to be the most cost-effective factory, and the suggestion was made to transfer more production to Illinois. The Holt Manufacturing Company was also carrying $5.6 million in short-term bank loans, with significant amounts coming due during 1924.[2]

With the Best manufacturing plant in San Leandro producing record volumes of Sixty and Thirty tractors, and sales at an all-time high, this favorable business situation allowed the focus of the Best executive committee to be centered on how best to attain control of the competitor. The financial aspects were turned over to Harry Fair. Fair, vice president of the investment banking firm of Peirce, Fair & Company of San Francisco, had been a member of the executive committee of the Best Company since 1919. He had devoted his time and expertise to establishing a solid financial base for the company. In 1924, the C.L. Best Tractor Co. showed earnings of 30 percent.[3] With such a strong showing in the previous year, Harry Fair commenced negotiations from a position of power.

One concern that had to be addressed was the tension that existed between the Best and the Holt companies. Since the turn of the century, the competition had been keen with numerous court battles and even a takeover of Best Manufacturing by Holt in 1908. C. L. Best resigned from his position with the Holt-controlled Best Manufacturing in 1910 and soon started his own company in Elmhurst. The two firms had been in direct head-to-head competition ever since. To make negotiations less awkward, C. L. Best was not directly involved. He chose to stay in the background and provide counsel while Harry Fair presided over the talks. Fair arrived at a solution that would allow the Holt Company to protect its reputation while still accomplishing what the Best consortium sought. If enacted, both the Holt and Best

1. E. E. Wickersham, unpublished manuscript (1940), Special Collections, University of California Library, Davis.

2. Reynold M. Wik, *Benjamin Holt & Caterpillar: Tracks & Combines* (St. Joseph, MI: American Society of Agricultural Engineers, 1984): 105.

3. "Merger of Best–Holt Tractors Eagerly Awaited," *San Francisco Examiner* (February 25, 1925): 20.

Best tractors painted in the new color scheme: gray and red. San Leandro Public Library Historical Photograph Collection #01489.

companies would cease to exist. A new third company, neither Holt nor Best, would take control of all of the operations. This new corporation would not be located in either San Leandro or Stockton, but in San Francisco.

The inside knowledge that Best plant manager Oscar Starr had of the Holt Peoria facility was important. During the Hawkins takeover period of 1917–18, Starr was wooed away from Best by Pliny Holt of the Holt Manufacturing Company by the offer of more money and the challenge to increase the production of war tractors at the Peoria plant. Starr also oversaw the installation of machinery in several new buildings. In mid-1918, Oscar Starr left the Holt organization and returned to San Leandro and the Best Company. Since the end of World War I, Holt had made few changes to its product line and to the Peoria production facility. While Best was developing and producing the model Sixty Tracklayer and later the model Thirty, Holt had done little in the ensuing six years to answer the Best challenge.

C. L. Best painted his Tracklayers in a distinctive color scheme: gleaming black with gold and red accents. The paint scheme of the Holt Caterpillars was more subdued. Anticipating the emergence of a new company and the difficult negotiations that would precede it, Best decided to paint the San Leandro tractors a color more palatable to the Holt concern. Dan G. Best recalls:

> *It must have been late '24 or early '25 that Dad decided the color of the Sixty needed to change. He talked to someone in the Navy yard in Oakland and found out they had a lot of surplus battleship gray paint. He got a good deal on it. The Sixty went to gray and that's how it stayed.*[4]

By early February 1925, the Pacific financial markets were alluding to "important developments" concerning the Best and Holt companies. The *Stockton Record* reported a meeting being held at the law offices of C. L. Neumiller, attorney for the Holt Manufacturing Company and a shareholder in the company. Murray M. Baker, vice president of the Holt Company and general manager of the Peoria plant, arrived in Stockton and participated in the negotiations. For several weeks, the Stockton area heard persistent reports that there "is something in the air" at the Holt plant.

On February 25, 1925, Oscar H. Fernbach, financial editor for *The San Francisco Examiner*, broke the story that had been simmering for quite some time:

> *San Francisco's men of affairs and of money—and indeed all of those throughout California—sitting in the theater of financial and industrial development are eagerly awaiting the curtain to rise upon an amalgamation of the C.L. Best Tractor Company and the Holt Manufacturing Company. Although most of the audience has been kept in the dark, there is no doubt that the stage is being set for the show.*
>
> *The utmost secrecy, the most profound silence, surround the preliminary transactions, which are known, however, to be going on. And when, at last, the time comes for the public to gaze upon the performance, it's dollars to doughnuts that the smaller of the actors will swallow the larger. In other words, signs point to the absorption, in some manner, of the Holt Manufacturing Company by that prince of money earners, the C.L. Best Tractor Company. Just how the combination is to be accomplished remains deep under cover. There is the Sherman law to be reckoned with.*
>
> *Harry Fair, of Peirce, Fair & Company, has "made" Best Tractor, so far as the financial accomplishment of that corporation is concerned. In conjunction with his efforts... stands the wondrous output of the corporation, especially in track-laying tractors. The Holt company, of course, is a big concern, but the showing made by the Best corporation... leaves it in the position to dictate terms. The combination is desirable from every point of view. That it is under way is best evidenced by the advance in Best stock.*[5]

This turn of events had the interest of many areas of California. San Leandro and Stockton, the respective homes of the

4. Personal interview with Daniel G. Best (April 6, 2006).

5. "Merger of Best–Holt Tractors Eagerly Awaited," *San Francisco Examiner* (February 25, 1925): 20.

Stop advertising when—
—NO. 2—
Strangers and travelers cease to visit Stockton.

Stockton Daily Morning Independent

WEATHER
All California: Fair and warm Saturday; light northerly winds.

64TH YEAR—VOL. 128—NO. 28. STOCKTON, CALIFORNIA, SATURDAY MORNING, FEBRUARY 28, 1925.

HOLT COMPANY CONTROL BOUGHT BY C. L. BEST TRACTOR HEADS

Best and Holt concerns, had much riding on the outcome of the negotiations. The San Francisco financial district also watched the events closely. To the north, the *Woodland Daily Democrat* kept local residents, many of whom had Best stock, informed of the progress of the talks. The groups didn't have long to wait. The headline of the February 28 Saturday morning *Stockton Independent* left little doubt of the outcome.

Harry H. Fair issued a short statement on Friday evening that ended some rumors but fueled others:

> *Peirce, Fair & Co., with a group of associates, today purchased controlling stock interest of the Holt Manufacturing Company. There is included in the group which has purchased this control practically all the executives of the C.L. Best Tractor Co. and members of the Holt family as well as the Benjamin Holt estate, which heretofore owned control of the company.*
>
> *No changes in the company, other than those that will in due course be dictated by good business judgment, are contemplated.*[6]

At the time it was suggested that much of the stock purchased by the group had been held by Thomas A. Baxter. Baxter had become president of the Holt Company when company founder Benjamin Holt died in 1920. He came to Holt as a business manager from a position with Bond & Goodwin, investment bankers with offices in San Francisco and New York City. Bond & Goodwin provided much of the working capital for the Holt concern. A short statement published in the February 28 edition of the *Stockton Record* had Baxter admitting that he had sold his interests in the Holt Company. He refused to comment on any reorganization of the company. On March 3, the *Oakland Tribune* reported that Charles L. Neumiller, an attorney from Stockton who had been actively involved with Best/Holt negotiations, had been elected president of the Holt Manufacturing Company. This was precipitated by the resignation of Thomas A. Baxter from the position he had held for five years.

In commenting on the announcement by Harry Fair, the *San Francisco Examiner*'s financial editor, Oscar Fernbach, said that this

> *gave the public confirmation of the first step in the operations whereby the Holt Manufacturing Company is to be absorbed by the C.L. Best Tractor Company. No one supposes that Peirce, Fair & Company, investment bankers, are going into the tractor business. The financial district*

6. "Harry H. Fair of C.L. Best Co. Purchases Controlling Interest in the Holt Manufacturing Co.," *Stockton Daily Record* (February 28, 1925): 1.

will wait with interest the announcement of the details under which the amalgamation is to be arranged.[7]

Those details were slow in coming. No further information was released to the newspapers. The ongoing talks were kept confidential. But while the discussions weren't in the public eye, a visit from the "tractor magnate" from San Leandro made headlines in the Tuesday, March 10 edition of the *Stockton Independent*:

Tractor Magnate Inspects Factory

Announcement Will Be Made When Plans Made

"The Holt plant will be kept in operation." This was the statement made to the Independent last night by C. L. Best, head of the C.L. Best Tractor Company of San Leandro, who spent the day in Stockton with a group of associates and officials of the Best Company. The tractor manufacturer would not authorize any further statement. He said: "We have not developed our plans, have not had time to decide any details and are just now busy inspecting the plant and planning for the future. We will have an important announcement for the city of Stockton just as soon as our plans are worked out." Best is accompanied by B.C. Heacock and Oscar L. Starr of San Leandro and R.C. Force of Oakland, officials of the Best company. . . . Announcement that Best and his associates had purchased a stock controlling interest in the Holt Company was made February 28.[8]

Stop advertising when— —NO. 7— You would rather fail than take advice and succeed.

Stockton Daily Independent Morning

54TH YEAR—VOL. 128—NO. 38. STOCKTON, CALIFORNIA, TUESDAY MORNING, MARCH 10, 1925.

BEST TO KEEP HOLT PLANT GOING

ROADS ACQUIESCE IN SUBWAY ORDER

VALLEY FRUIT SHIPMENT FRAUD CHARGED

STOCKTON TO OPPOSE CLOSING OF

Doheny Companies Are Sued For $904,551 Income Taxes Due 1919 On Over $2,000,000

HOTALING CHARGES

Under the Dome At Sacramento

$225,000 NETTED BY

Tractor Magnate Inspects Factory; Announcement Will Be Made When Plans Made

Harry Fair was the only one of the five-member Best management team who did not make the journey to Stockton. Fair had gone east on a prolonged visit.[9] As mentioned in the *San Francisco Examiner* article from February 25, the Sherman Antitrust Act could have been a factor in the takeover. The Sherman Act, passed by Congress in 1890, sought to prohibit abusive monopolies. An abusive monopoly is described as one that engages in any of a variety of anticompetitive practices. In a historic decision, the Supreme Court stated that

large size and monopoly in themselves are not necessarily bad and do not violate the Sherman Antitrust Act. Rather, it is the use of certain tactics to attain or preserve such position that is illegal.[10]

With Harry Fair in the East dealing with possible difficulties, changes happening at the Holt Stockton plant were reported by the *Oakland Tribune*:

STOCKTON, March 11—The first effect of the consolidation of the Holt Manufacturing Company and the Best Tractor Company was seen here Monday [March 9], when more than fifty employees of the old Holt Company were discharged. Operations here have been curtailed. C.L. Neumiller, new president of the [Stockton] Best Company announced yesterday he could make no statement at present, except to say the Holt plant would remain here and that eventually a bigger and better establishment would be built up.[11]

7. "Best Tractor," *San Francisco Examiner* (February 28, 1925): 21.

8. "Tractor Magnate Inspects Factory," *Stockton Independent* (March 10, 1925): 1.

9. "Best–Holt Merger is Indicated," *Woodland Daily Democrat* (April 15, 1925): 1.

10. "The Sherman Anti-Trust Act," Linux Information Project (http://www.linfo.org/sherman.html).

11. "Work Reduced in Plant Consolidation," *Oakland Tribune* (March 11, 1925): 3.

By mid-April, Harry H. Fair had returned to San Francisco and rumors again started in the financial district concerning the Best and Holt companies. The stock market reacted with

> *Best stock skyrocketing to new heights and Holt performing in a somewhat similar manner. . . . Inquiries at the offices of Peirce, Fair & Co., as to the reported merger, and to any announcement being made were barren of results. Officials of the company refused to either affirm or deny the reports of the linking together of the two companies. Newspaper men seeking details of the reported merger were told that they might draw their own conclusions, and say what they wished. This was entirely in line with the attitude of Peirce, Fair & Co., recently, when the concern acquired control of Holt. After refusing to either confirm or deny the purchase of Holt, the street, within twenty-four hours, received formal announcement of the deal.*[12]

While the newspapers were wondering about the fates of the two companies, legal proceedings that would alter the tractor industry were completed in San Francisco. As announced by Peirce, Fair & Company in an official statement, the Caterpillar Tractor Co., headquartered in San Francisco, was incorporated under California laws on April 15, 1925. That corporation was formed to

> *acquire all of the assets, patents and trade-marks of the C.L. Best Tractor Co., a California corporation heretofore the sole manufacturer of BEST "TRACKLAYER" TRACTORS, and of the Holt Manufacturing Company, another, California corporation, heretofore the sole manufacturer of HOLT "CATERPILLAR" TRACTORS.*[13]

Many people were eager to hear the details of Caterpillar's acquisitions from Best and Holt, as well as where the governing power would rest. The announcement of the board of directors

12. "Best–Holt Merger Indicated," *Woodland Daily Democrat* (April 15, 1925): 1.

13. "C.L. Best of San Leandro to Head New Corporation," *Oakland Tribune* (April 17, 1925): 35.

Preliminary Proof

260,000 SHARES

CATERPILLAR TRACTOR CO.

CAPITAL STOCK

This corporation is being formed under the laws of the State of California to acquire all of the assets, patents and trade-marks of the C. L. Best Tractor Co., a California corporation heretofore the sole manufacturer of BEST "TRACKLAYER" TRACTORS, and of the Holt Manufacturing Company, another, California corporation, heretofore the sole manufacturer of HOLT "CATERPILLAR" TRACTORS.

The new company will own the following plants:

(a) San Leandro, California, a plant occupying 10 acres of which 5½ acres are under roof, fully equipped for the complete, economical manufacture of 2500 "Tracklayer" Tractors per year.

(b) Stockton, California, a plant occupying 21 acres of which 12 acres are under roof, equipped for the manufacture of combined harvesting machines and for the manufacture of spare parts for "Caterpillar" Tractors.

(c) Peoria, Illinois, a plant occupying 40 acres of land of which 14 acres are under roof, fully equipped for the manufacture of 5000 "Caterpillar" Tractors per year.

All of these properties are owned in fee and are unencumbered except that certain buildings (the values of which are not included in plant figures on the appended balance sheet) in the manufacturing plant group at Peoria, which were erected during the War by the United States Government, are held under favorable purchase agreement.

HISTORY

The C. L. Best Tractor Co. was formed in 1910 for the purpose of manufacturing gas engine propelled tractors. The Holt Manufacturing Company was established in 1883 as a manufacturer of farm implements. About fifteen years ago both companies began the development of the track laying type of tractor and in the intervening years have brought the product to a high state of mechanical efficiency. Most of the important patents in connection with the art have been issued to or acquired by one or the other of these companies and will become the property of the new Company.

PRODUCT

"Caterpillar" and "Tracklayer" tractors have become standard implements in many important industrial activities, notably in the fields of road building and maintenance, excavating, grading, logging, snow removal and agriculture where a large, mobile traction power plant is necessary. During the period since the War the utility of these machines has broadened and is broadening very rapidly. The product is sold throughout the world. The Stockton plant, aside from the manufacture of caterpillar tractors and parts, has for many years manufactured "Holt" combined harvesters which rank high in that field.

SALES

The combined sales of the two constituent companies for the year 1924 were approximately $17,500,000. Trade indications at the beginning of the year pointed to a largely increased volume of tractor sales in 1925, and sales to date for this year amply confirm this indication.

EARNINGS

For the year 1924 the Best Company's earnings, before Federal income taxes, were $901,000 on a sales volume of $5,300,000. Similar earnings for the previous year were $703,000 on a sales volume of $4,000,000. Sales volume for the Holt Company for each of the years 1923 and 1924 was in excess of $12,000,000. The executive management of the new Company will be largely in the hands of the men who have been responsible for the success of the Best Company. Economies in the Holt manufacturing operations already effected justify the prediction of earnings for the new Company, before Federal income taxes, for the year 1925 in excess of $2,250,000. Savings to be gradually realized from purchasing, manufacturing and sales co-ordination should substantially increase net earnings in 1926, without allowance for increased volume of sales, which can be confidently expected.

The board of directors and official staff of the Catepillar Tractor Co. will consist of

C. L. Best, Chairman of the Board
R. C. Force, President
B. C. Heacock, Vice Pres't and Secretary
M. M. Baker, Vice President
P. E. Holt, Vice President
O. L. Starr, General Factory Manager
Allen L. Chickering
Harry H. Fair
John A. McGregor

CAPITALIZATION

The capital structure of the new Company will consist of a single class of stock, of $25 par value, of which there will be presently issued 260,000 shares, out of an authorized total of 500,000 shares. Book value of the stock to be presently issued, on the basis of the appended combined balance sheet, will be approximately $50 per share. However, the depreciated reproduction value of plants exceeds book value by approximately $1,500,000, or more than $5 per share.

DIVIDENDS

It is contemplated that dividends will be established at the rate of $1.25 per share per quarter.

WORKING CAPITAL

The quick assets of the new Company, after giving effect to the financing in process, as shown on the appended combined balance sheet, as of December 31, 1924, totaled $11,974,027. The floating debt amounted to $4,182,242. This is a current asset ratio of approximately three to one. It is believed by the management of the new Company that, as a result of economies incident to unified operations, cash can be extracted from receivables and inventory sufficient to materially redrce if not extinguish this floating indebtedness within one year.

Preliminary proof. Caterpillar Tractor Co. Courtesy of Dan Best II.

and official staff of the Caterpillar Tractor Co. left little doubt of just who was in control. In an article in the April 17 *Oakland Tribune,* it was reported that the executive management "of the Caterpillar Tractor Company will be largely in the hands of the men who have been responsible for the success of the Best Company."[14] The members were listed as:

C. L. BEST, CHAIRMAN OF THE BOARD
R. C. FORCE, PRESIDENT
B. C. HEACOCK, VICE PRESIDENT AND SECRETARY
M. M. BAKER, VICE PRESIDENT
P. E. HOLT, VICE PRESIDENT
O. L. STARR, GENERAL FACTORY MANAGER
ALLEN L. CHICKERING
HARRY H. FAIR
JOHN A. MCGREGOR

The former positions held by the new directors and staff were explained in an article on May 8, 1925, in the *Oakland Tribune*:

C. L. Best was formerly president of the C.L. Best Tractor Company; R. C. Force has been for many years vice-president and general manager of the C.L. Best Tractor Company; B. C. Heacock was an executive of the C.L. Best Tractor Company and M. M. Baker was formerly vice-president in charge of the Peoria plant of the Holt Manufacturing Company. P. E. Holt was engineering vice-president of the Holt Manufacturing Company. O. L. Starr was factory manager of the C.L. Best Tractor Company. Harry H. Fair has been for many years a member of the executive committee of the Best Tractor Company, is vice-president of Peirce, Fair & Co. and has been largely instrumental in the consolidation of the companies. Allen L. Chickering is a San Francisco attorney of the firm Chickering and Gregory. John A. McGregor was formerly president of the Union Iron Works, now the Bethlehem Shipbuilding Corporation.[15]

Of the nine men listed, only two had ties with the Holt Company: Murray Baker and Pliny Holt. Allen Chickering's law firm represented Peirce, Fair & Co., and John McGregor was the owner of the firm that hired Oscar Starr as an apprentice machinist early in Starr's career. The group of five "Right Men"—Best, Fair, Force, Heacock, and Starr—which had been together in the C.L. Best Tractor Co. since 1919, continued together in the Caterpillar Tractor Co. and would remain together for over twenty-five years.

With the change in management, the Caterpillar Tractor Co. acquired ownership of the three manufacturing plants:

San Leandro, occupying 10 acres of which 5 and a half are under roof, fully equipped for the complete, economical manufacture of 2500 "Tracklayer" Tractors per year [Sixty and Thirty models]; one at Stockton, occupying 21 acres of which 12 are under roof, equipped for the manufacture of combined harvesting machines and for the manufacture of spare parts for "Caterpillar" Tractors; and another at Peoria, Illinois, occupying 40 acres of land, of which 14 acres are under roof, fully equipped for the manufacture of 5,000 "Caterpillar" Tractors per year [Ten Ton, Five Ton, and Two Ton models].[16]

The next step in the consolidation process was for stockholders of each company to meet and to agree to the terms offered. The Holt Manufacturing Company stockholders met on May 4 in Stockton. The C.L. Best Tractor Co. stockholders held a similar meeting the following day, May 5, in San Leandro. The stockholders of both companies ratified the actions of their respective boards of directors, thus allowing for the exchange of the assets of each company for stock in the newly

14. Ibid.

15. "Best and Holt Cease Official Life Next Week," *Oakland Tribune* (May 8, 1925): 47.

16. "Caterpillar Tractor Co. Takes Over Interests of Best–Holt Companies," *San Leandro Reporter* (April 24, 1925): 1.

formed Caterpillar Tractor Co. On Thursday, May 14, 1925, the C.L. Best Tractor Co. and the Holt Manufacturing Company ceased independent operations. All plants and businesses were operated by and all assets and liabilities assumed by the Caterpillar Tractor Co.

Call it a merger, a consolidation, or an amalgamation, but the dominance of C. L. Best in the track type tractor field at the time was affirmed on May 14 with Best people in the positions of power and Best products and Best manufacturing principles set to carry the new company into the future.

As Daniel G. Best, son of C. L. Best remembered:

> *Dad always said that if they'd waited a bit longer, they could have bought Holt on courthouse steps in Stockton, but they didn't want it to get away. He wanted that name. Dad wanted that name Caterpillar.*[17]

Best Sixty Tracklayer recovering a Holt 10 Ton Caterpillar. The Sixty continued in production until 1931. The 10 Ton ceased production shortly after the incorporation of the Caterpillar Tractor Co. in 1925. Special Collections, University of California Library, Davis.

17. Personal interview with Daniel G. Best (April 6, 2006).

Caterpillar Thirty tractor. Photo courtesy of Col. O.P. Winningstad, U.S. Army.

The New Era

Friday, May 15, 1925, dawned with the employees at the Caterpillar plants in Peoria, San Leandro, and Stockton starting an ordinary workday, but the events of the previous day would bring changes that few of the plant workers could imagine. The operations at the Peoria and Stockton plants were scrutinized: Everything was on the table. The management at Caterpillar made the decisions that would be in the best interests for the future of the new company. President R. C. Force, in speaking about the transition from two companies to one:

> *[It] makes available under one control the research and promotion departments of both companies. It brings to us men of varied experience and capabilities. It has been pleasant to be able to lay down the experience and talents of one man beside those of another and to place these men in positions for which they are best adapted. Few eliminations have been necessary. Without confusion we have blended the two organizations, and have coordinated the work so as to obtain the highest efficiency at the various plants.*[1]

The San Leandro plant kept producing the Sixty and Thirty models, now Caterpillars. The first tractor made after the formation of the new company was a Model Sixty tractor, serial number 2547A with patent plate part number 2777A. The 2 Ton, 5 Ton, and 10 Ton Caterpillars were kept in production at Peoria while that plant was being geared up to produce the Sixty and Thirty models. Heat treatment of parts, long a staple with Best products, had to be improved and expanded at Peoria. The five tractor models offered by Caterpillar were now painted gray with red lettering and featured the trademarked red wavy line. By the end of November 1925, the 10 Ton had been dropped from production. Likewise,

1. F. Hal Higgins Collection, Special Collections, University of California Library, Davis.

(Above) *Caterpillar tractor school hosted by the Wm. H. Ziegler Co. Inc. in Minneapolis, Minnesota. Ziegler became a Best dealership in 1919.* Photo courtesy of Ziegler, Inc. (Left) *Caterpillar school graduates. Note the logo on the coveralls.* Photo courtesy of Ziegler, Inc.

the harvester production at Stockton was halted and the plant shut down. As reported to Oscar Starr on September 9, 1925:

> *There are approximately eighty men on the factory pay-roll to date. This will hold good for about three weeks, after which it will drop to fifty-five men, as the Harvesters and a number of rebuilt '45' Tractors, on which we are doing some work to make them saleable, will be finished. The scrapping program is about finished; that is the Tractor parts. We expect the Harvester scrapping to finish about October 15th.*[2]

The Caterpillar Tractor Co., like its predecessors, relied on a network of independently owned domestic dealerships to provide parts and service for the machines they sold. With many overlapping territories, difficult decisions had to be made as to who would be retained. By the end of 1925, eighty-nine dealers were set to officially represent the Caterpillar Tractor Co.[3] I. E. Jones, sales manager for the tractor company, while in Billings, Montana, on his annual trip to Western dealerships, expounded on the importance of cooperation between the manufacturer and the dealers. He noted the company intended to "furnish better tractors for less money than ever before." Jones further said that the company's dealership plan would put parts and service in the hands of every dealer and that parts prices had been considerably reduced due to the company's extended field of operations.[4]

One idea retained from the Holt Manufacturing Company was the "Caterpillar School." These schools provided training for both dealer servicemen and customers. The schools proved both popular and successful. The new company embraced the idea and expanded upon it.

When the company was first incorporated, the main office of the Caterpillar Tractor Co. was located in San Francisco. By

Caterpillar corporate headquarters, 800 Davis Street, San Leandro, California. Photo courtesy of Col. O.P. Winningstad, U.S. Army.

mid-July 1925, that office was moved to San Leandro. In making that announcement, Caterpillar President R. C. Force also noted that new larger offices were being contemplated and that the business was running smoothly, with the factory departments running at full capacity.[5]

With the San Leandro and Peoria plants booming, the Stockton harvester facility was closed. This was a blow to the economy of the Stockton area. When, on December 29, 1925, the announcement was made by Harry H. Fair that the plant would reopen, the citizens of Stockton were delighted. The old Holt

2. Letter from E. J. Weber, general superintendent, to Oscar Starr, Pliny Holt papers, The Haggin Museum, Stockton, California.

3. "Best–Holt Merger Brings Policies to Guide Caterpillar's Future," *Caterpillar World: Century of Change* (May 1984): 26.

4. "Evolution of Caterpillar Greater Than That of Auto Declare Big Tractor Man," *Billings Gazette* (April 25, 1926): 8.

5. "Main Office of Caterpillar Tractor Company to Be Located in San Leandro," *San Leandro Reporter* (July 17, 1925): 1.

harvester plant would now house the newly formed Western Harvester Company. This company

> *would be capitalized for $2,000,000, with all the stock held by the Best Caterpillar Company, but [would function] as an entirely separate entity, with its own board of directors and executive staff.*[6]

The management staff for the Western Harvester Company was announced as A. S. Weaver, president and general manager; B. C. Heacock, vice president and secretary; L. H. Thoen, factory manager; and T. H. Luke, sales manager. The Holt Combined Harvester was well known both in the United States and abroad and the new company intended to tap into that notoriety. Production of harvesters at the Stockton plant continued through the remainder of the 1920s. By 1930, the Caterpillar harvester was manufactured at the Peoria facility.

Total Sales of Caterpillar Were Near 21 Million

Debts of 5½ Million Paid Off Last Year; Outlook Is Bright[7]

That headline and the accompanying story in the *Oakland Tribune* of February 20, 1926, reported the excellent news that stockholders had received in the 1925 Caterpillar Tractor Co. annual report. The $21 million figure represented a 19 percent increase over the combined sales of the Best and Holt companies in 1924. Total earnings, before federal taxes and interest, were just under $4.5 million. When the company was formed in May, it had assumed the liabilities of the two predecessors. The total of outstanding notes and accounts payable was $5.5 million. At the end of 1925, the company had paid those debts and only owed federal taxes and current accounts payable amounting to less than $1 million. Also at the end of 1925, statistics

6. "Holt Plant Becomes Scene of Great Harvester Industry," *Stockton Daily Independent* (December 29, 1925).

7. "Total Sales of Caterpillar Were Near $26 Million," *Oakland Tribune* (February 20, 1926): 11.

HERE IS NEWS
of the utmost
importance to every
grain grower

ANNOUNCEMENT

The Western Harvester Company has been formed to continue the development of what has grown to be one of the most important and successful Combined Harvester businesses of the country—a business brought by the Holt Manufacturing Company during the past 40 years, to the point where further development demands this new, separate organization.

The Western Harvester Company will manufacture the "Holt" Models 30 and 32 Steel Harvesters in the former Holt plant at Stockton. It will supply parts for former models of "Holt" Harvesters. It includes in its organization the men chiefly responsible for the successful design and construction and sale of "Holt" Harvesters in the past. It will sell its product through an organization of reliable, eager-to-serve dealers.

The entire capital stock of this new company will be owned by the Caterpillar Tractor Co. Concentrating its efforts, its manufacturing facilities and its organization on the production of combined harvesters, Western Harvester Company offers a better product, better service, better user-satisfaction than have ever before been available to harvester purchasers.

Write for "Holt"
Combined Harvester
Literature

WESTERN HARVESTER COMPANY
General Offices and Factory, Stockton, Calif.
Distributing Warehouse, Spokane, Wash.
Manufacturers of "Holt" Combined Harvesters

Newspaper advertisement announcing the forming of the Western Harvester Company. Woodland Daily Democrat, *January 6, 1926.*

"Running in" a new Caterpillar Twenty engine. San Leandro, California. Photo courtesy of Col. O.P. Winningstad, U.S. Army.

Caterpillar Sixty providing belt power for a portable sawmill on a farm near Howard Lake, Minnesota. 1942. Author's collection.

showed that the Caterpillar Tractor Co. had a total of 2,528 employees and had paid $2.9 million in wages, salaries, and employee benefits.

The next four years saw phenomenal growth for the company. Sales peaked at $51 million in 1929.[8] New tractors were in development. The first truly Caterpillar Tractor Co. machine was the Twenty. This 20 H.P. tractor was designed in San Leandro. Production began in 1927 in San Leandro and the following year in Peoria. This tractor, like the Sixty and Thirty models, continued the Best tradition of a model letter designation. The Sixty was the Model A, the Thirty was the Model S, and the new Twenty was given the letter L. When the tractors were produced at the Peoria plant, the letter designations stayed the same, but were preceded by the letter P: PA, PS, and PL. The decade was rounded out with two additional models going into production: the Ten, or Model T, was a 10 H.P. tractor introduced in 1928 and the Fifteen, or Model V, introduced the following year.

While the introduction of new tractor models was a highlight of the late 1920s, a much more fundamental change also took place. In his patent dated March 19, 1929, C. L. Best stated: "In experimenting with and developing tractors over a long period I have found that one of the most important features affecting the commercial success of a track type tractor is the track shoe." Best's invention called for the track shoes to be made from a metal bar that had been rolled to the width and form of the shoe, but in a long length that allowed multiple shoes to be sheared from it. The holes required for fastening would then be punched into the individual shoes. Best claimed that these shoes "are of excellent quality primarily due to the rolling process, and are quickly and economically produced."[9] This rolling process became the industry standard and is still in use today.

As the uses for the different sizes of Caterpillar tractors increased, other companies offered allied equipment designed to work with those machines. Caterpillar dealers sold that equipment as a service to their customers and to increase their profits. When the decision was made to go after the sales of road-building equipment, the company looked to purchase a line rather than designing it from the beginning. On December 4, 1928, the Caterpillar Tractor Co. purchased the Russell Grader Mfg. Co. This Minneapolis, Minnesota, concern was well known for the quality and diversity of the road-building equipment it produced. The first catalog after the purchase announced the

8. "Depression Era Sales Helped by Large Orders from USSR," *Caterpillar World: Century of Change* (May 1984): 29.

9. "Method of Providing Track Shoes," U.S. Patent No. 1,705,802. C.L. Best assigned to Caterpillar Tractor Co. (March 19, 1929): 1.

merger of the Caterpillar Tractor Co. and the Russell Grader Mfg. Co. In order to better serve the public, the above named companies have been merged into one unit. Their combined resources and facilities make it possible to develop a better product more economically and to render a more efficient service.[10]

Dealers now had the advantage of transacting business with one company. This streamlined sales and the availability of parts and service. The future looked bright for the new Road Machinery Division of the Caterpillar Tractor Co.

Even while the company was still adjusting to changes wrought on May 14, 1925, C. L. Best envisioned what the next ten years could hold for the Caterpillar Tractor Co. That future had its roots in a visit by Best and Oscar Starr to the 1915 Pan Pacific Exposition held in San Francisco. The Palace of Machin-

(Above) *The former Russell Grader Mfg. Co. headquarters in Minneapolis, Minnesota. The building has been renamed the Caterpillar Tractor Co. Road Machinery Division. 1930.* Photo courtesy of Ted Knack. (Top) *Albert Knack, employee of the Caterpillar Tractor Co. Road Machinery Division, driving a Caterpillar 2 Ton tractor moving a Model Sixty grader. 1932.* Photo courtesy of Ted Knack.

10. "Russell Road Machinery," Caterpillar Tractor Co. (1929): 1.

ery showcased many innovations. One that caught the attention of Best and Starr was a Danish-built stationary Diesel engine powering a generator. According to Oscar Starr, "Leo and I used to go over and look at it. 'We got to have it some day,' we agreed. On up into the 1920s, we were always interested and kept posted on diesels over the world."[11] That someday waited while C. L. Best put his Sixty into production, solved the issues inherent with track reliability of tracklaying tractors, and acquired an eastern manufacturing facility. After 1925, some ten years after the "We got to have it someday" comment, research and development started in earnest. The Caterpillar Tractor Co. would be defined in the 1930s, 1940s, and beyond by one word: Diesel.

Caterpillar Fifteen Motor Patrol. 525 patrols were produced. Selling price for the tractor was $1,450 with the grader unit priced at $1,000. Sales brochure. Author's collection.

11. Oscar Starr interview, F. Hal Higgins Collection, Special Collections, University of California Library, Davis.

Caterpillar Tractor Co. San Leandro factory. 1935. Company brochure. Author's collection.

When the Diesel is Ready

Rudolf Diesel received a patent in 1892 for the compression ignition engine that would bear his name. At the time, this was one of many attempts by inventors to build a functional internal combustion engine. The ignition system—the means of igniting the fuel—was a problem that had inventors trying many solutions. Diesel's engine, using compression ignition, was more reliable than others built at that time.

By 1900, Diesel-type engines were being used commercially as stationary power units and as power for marine applications. These engines were slow speed and had a high weight-to-horsepower ratio. They did their assigned tasks well. But could the Diesel-type engine be used in a tractor? C. L. Best believed that just because a successful Diesel tractor engine had not been built didn't mean that it couldn't be built. He stood fast against skeptics in the company who felt that the current gasoline-powered models would continue to meet the demands of the marketplace in the future. Best had always been well served by embracing modern technology, and he was convinced that a Diesel-powered tractor held the key to the future of the Caterpillar Tractor Co.

Before the Caterpillar engineers began their research into the Diesel engine, clear goals had to be established. As stated in the company publication, *The Caterpillar Diesel*:

> *A Diesel engine for the Caterpillar track-type Tractor must be dependable. It must be simple, too, if the full opportunity of usefulness was to be realized. So simple that no special skill would be required for its operation and maintenance, so simple that the engine would be successful in the*

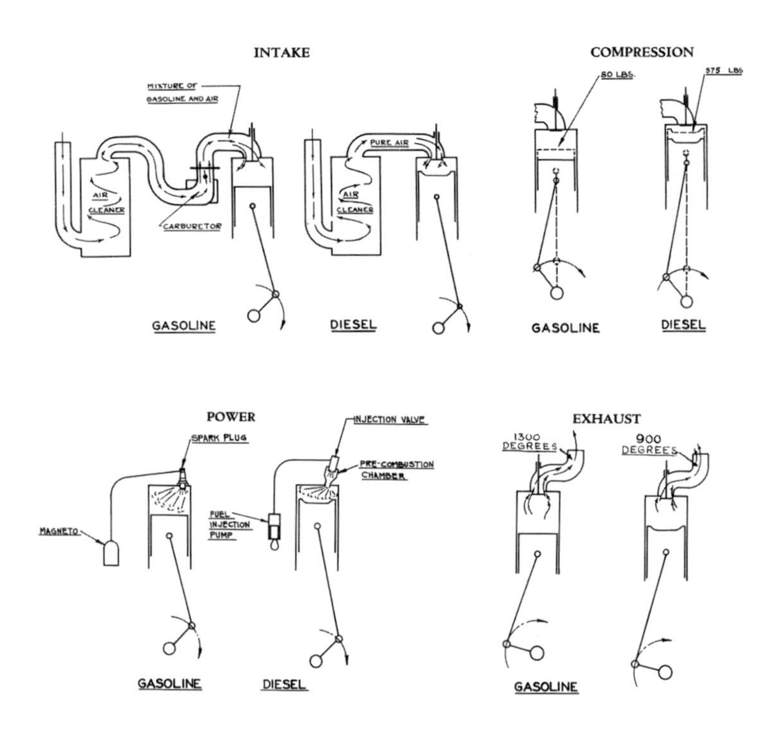

(Left) *These diagrams illustrate the operational differences of a gasoline and a diesel engine.* Company sales brochure. Author's collection. (Above) *A stationary 10-cylinder Diesel engine that develops 11,000 horsepower. This style of engine was designed to run at a constant speed.* Company catalog. Author's collection.

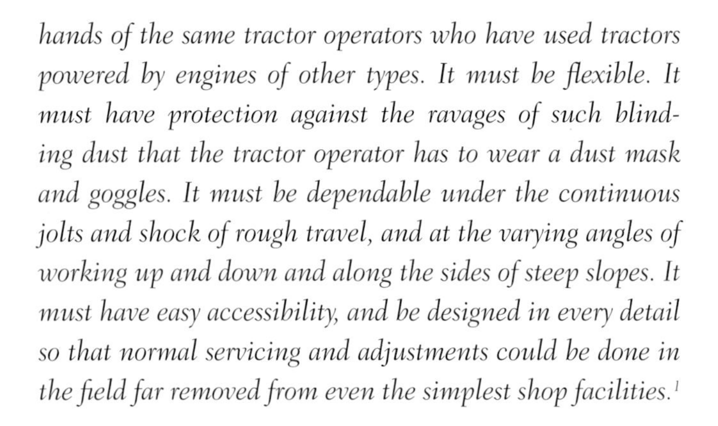

hands of the same tractor operators who have used tractors powered by engines of other types. It must be flexible. It must have protection against the ravages of such blinding dust that the tractor operator has to wear a dust mask and goggles. It must be dependable under the continuous jolts and shock of rough travel, and at the varying angles of working up and down and along the sides of steep slopes. It must have easy accessibility, and be designed in every detail so that normal servicing and adjustments could be done in the field far removed from even the simplest shop facilities.[1]

With those goals always in mind, the engineers at the San Leandro plant began the methodical testing of existing Diesel engines produced in the United States and abroad. These engines were purchased by the company and results from the testing showed no Diesel engine on the market met the requirements set forth by Caterpillar management. The test results left no alternative: The Caterpillar Tractor Co. would commit the money, time, and man power to design and build a Diesel engine of its own. As stated by Chairman C. L. Best, "When the Diesel tractor is ready, Caterpillar will have it."

The engineering staff at the San Leandro plant was well versed in the concept and design of the gasoline tractor, but an engineer specializing in Diesel design was needed. As mentioned earlier, Diesel engines were in use for various stationary and marine applications. One of the first to build the Diesel engine in the United States was George A. Dow. In 1911, Dow was in Europe and met with Dr. Rudolf Diesel and was shown around a number of manufacturing plants. He later established the Dow Diesel Pump and Engine Company, located in Alameda, California. This company built a total of twenty-eight engines. By 1926, after manufacturing Diesel engines for nearly fifteen years, Dow was ready to move on. In an interview given on the fortieth anniversary of Dr. Diesel's invention, Dow remembered:

Leo Best is an old schoolmate of mine. I recall seeing him at the Sequoia Club one day. It must have been along about 1926, and he asked me where he could get a high

1. "Qualities Demanded by Caterpillar," *The Caterpillar Diesel* (1933): 6.

Various testing equipment designed and used by Caterpillar engineers in the development of the Caterpillar Diesel. Company catalog. Author's collection.

grade Diesel engineer as he was going into Diesel tractor building. I said he could have my engineer, as I was ready to quit. That man was Rosen, whom I got from the University of California.[2]

Carl G. A. Rosen oversaw the development and testing of the Caterpillar Diesel engine. Not only did Rosen and his staff work on designing the engine, but they also had to design the testing equipment. The search for a suitable fuel-injection and combustion system led to the development of a single-cylinder engine known in the laboratory as "the rubber engine." This engine was not made from rubber but could be easily configured or "stretched" into nearly unlimited configurations as demanded by the researchers. The San Leandro engineering research department designed equipment capable of measuring the pressure and temperature within the cylinder, the amount of fuel consumed during the test, the amount of air being inducted, and the temperature of the exhaust gasses. A heater kept the air temperature constant during the testing. The progress was slow, and minute details had to be recorded so accurate comparisons

2. "Diesel Cuts a 40-Candle Cake," *Pacific Rural Press* (December 26, 1936): 707, F. Hal Higgins Collection, Special Collections, University of California Library, Davis.

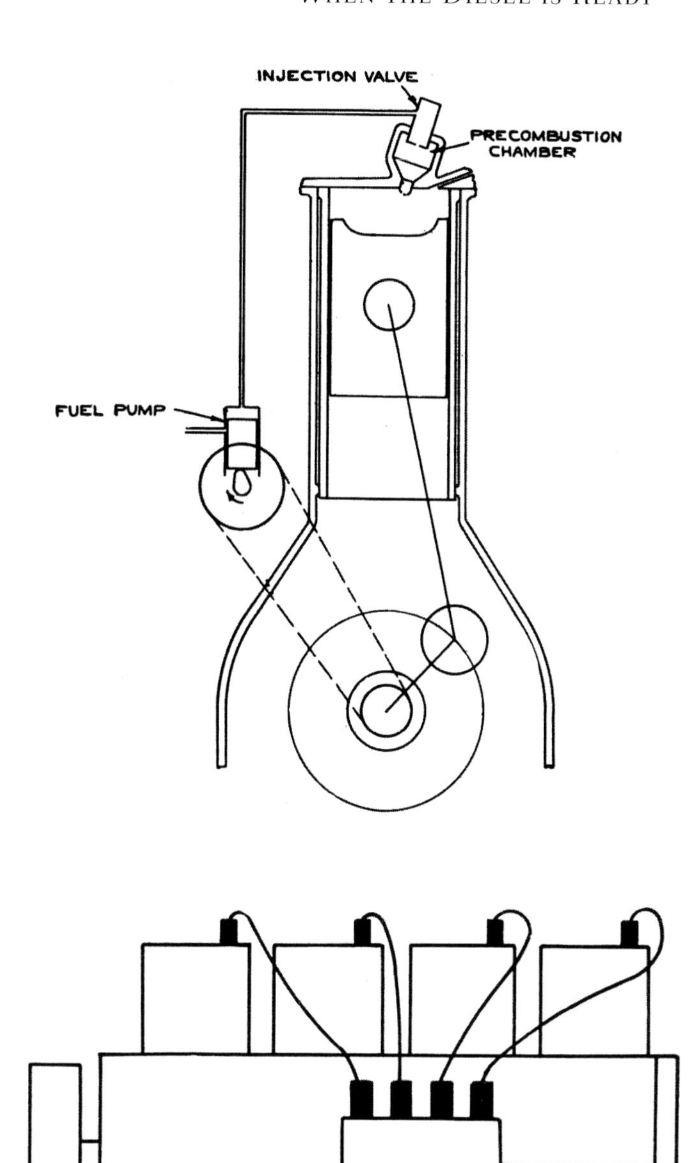

(Top) *This diagram shows the basic combustion concept of the Caterpillar Diesel.* Company catalog. Author's collection. (Above) *The fuel pumps on a Caterpillar Diesel engine were designed for complete interchangeability and required no calibration in the field.* Company catalog. Author's collection.

could be made. Time had to be split into thousandths of a second. Fuel spray that lasted one hundredth of a second was measured at a volume of 3 millionths (.0000003) of a gallon.[3] With initial phases of testing completed, C. L. Best and his engineers decided upon the basic design of the Caterpillar Diesel engine: It would be a four-cycle, solid injection engine employing pre-combustion chambers with individual fuel-injection pumps and would be started by a small gasoline engine.

The pace of the research and development increased in 1927. As stated by Warren G. Brown, a Caterpillar research engineer, in an interview conducted by F. Hal Higgins: "Out here in the West, California in particular, the incentive [is] to do the impossible. In this kind of research work, you don't alibi. You meet the problem and solve it or get off the dime."[4] And they continued to solve the engineering problems. But October 29, 1929, brought another set of problems that even C. L. Best, Oscar Starr, Carl Rosen, and their staff were unable to solve: the stock market crash on Black Tuesday and the start of the Great Depression.

As the domestic market for Caterpillar equipment withered, tractor sales to the Russian Grain Trust helped keep the company profitable. The Soviets were developing fifty-four large dormant acreages into fields for agricultural use. During November 1930, the Russians took shipment of a $5 million order that translated into 1,500 tractors. Sales to Russia in 1930 were over $10 million, or nearly 22 percent of the total sales for the Caterpillar Tractor Co.[5] The company showed a profit for 1929 of $12.3 million. The figure for 1930 was lower—$9.1 million. The year 1931 saw the profit drop lower still to $1.5 million, while 1932 was a disastrous year with a posted loss of $1.6 million.[6] But during this period of diminishing profits, the development of the Caterpillar Diesel engine never slowed. C. L. Best believed

"Old Betsy", serial number 1A14, was the first in a long line of Caterpillar Diesel engines. The engine was designed and built at San Leandro factory. "Old Betsy" was later donated by the Caterpillar Tractor Co. to the Smithsonian Institution. Company sales brochure. Author's collection.

that Diesel was the motive power of the future, and he never backed down and never thought of quitting. The company sank nearly $1 million into the project. The result of all the hard work and dedication appeared in June 1930: Engine serial number 1A14, nicknamed "Old Betsy" by the San Leandro employees, was the first Caterpillar-designed and -built Diesel engine.

With the engine marking the beginning of a manufacturing era, a letter to dealers, dated July 3, 1930, announced the end of another. The San Leandro plant would no longer manufacture tractors. Manufacture would be concentrated at the larger East Peoria facility. When the changes were complete, by the end of 1930, San Leandro would officially become the Research Division of the company. Oscar Starr, who had been in charge of manufacturing, would now be in direct charge of the laboratories and personnel. R. C. Force, the first president of the Caterpillar Tractor Co. was succeeded in that post by B. C. Heacock. The corporate home office would be continued in San Leandro, but

3. "The Combustion Problem," *The Caterpillar Diesel* (1933): 12.

4. Warren G. Brown interview, F. Hal Higgins Collection, Special Collections, University of California Library, Davis.

5. "Caterpillar Again Shipping to Russia," *Oakland Tribune* (November 17, 1930): 20.

6. "Caterpillar Has Grown Up, Too," *Caterpillar News and Views* (Sept./Oct. 1954): 16–17.

To make fuel savings of the Diesel engine available to owners of a gasoline-powered Sixty tractor, Caterpillar offered a conversion kit that allowed the Diesel engine to be mounted in a Sixty running gear. Author's Collection.

the members of the operating staff would now call Peoria home. Force stated that the new operating plan had been well thought through and would provide "the necessary close supervision over the increasingly large operations of the Company at its major plant. . . ." Four out of the original 1919 five "Right Men" (Best, Fair, Force, and Starr) decided that a move to Peoria would not be for them. As told by C. L.'s son Dan, "they all had to take a pay cut to stay in California. Dad did take the train back to Peoria when he had to, but he was always glad to get back to California."[7] After the end of tractor production

> *at San Leandro it is planned to bring together a corps of engineers from the various divisions of the Company to work in connection with the Chairman of the Board, C.L. Best, whose inventive genius has contributed so tremendously to the present high development of tractors and other "Caterpillar" products.*[8]

Testing continued on Old Betsy. Using the information gathered from over one year of further testing on the prototype, two engines were built. The 4-cylinder, 6⅛" bore, 9" stroke engine was designed to fit into the running gear of the Model Sixty tractor. Due to the increased weight and torque of the Diesel engine, the main frame of the tractor was strengthened and the equalizer bar was replaced with a more heavy-duty equalizer spring. In August 1931, the shipping weight of the new Diesel tractor was approximately 25,000 pounds, while the gasoline-powered Sixty came in at 20,500 pounds. This marriage of engine to running gear gave the Caterpillar Tractor Co. its first Diesel-powered tractor.

In a letter to dealers dated August 31, 1931, the company proudly announced the Caterpillar Diesel Sixty Tractor. It was stressed that it was Caterpillar designed and built and resulted from years of experiments and testing under many conditions. The new Caterpillar Diesel Sixty tractor offered

- the satisfaction of obtaining the lowest possible fuel cost
- the feel of unconquerable power, quickly responsive, steady under widely varying loads, and confident power
- the new high standard of Diesel performance—simplicity, standardization of parts, and accuracy without intricacy
- a Diesel engine designed for tractor service-rugged, dependable-built to meet field conditions.[9]

The letter went on to emphasize how production of these tractors was quite limited and how "only a few Caterpillar dealers will be privileged to make deliveries of these machines to their customers." As these were the first successful Diesel-powered tractors, the research division insisted on close observation in the field. Owners of the tractors would be requested to keep careful records of fuel usage and cost of repairs and "provide a full schedule of work to keep the tractors busy, preferably on both day and night shifts."[10]

The first Caterpillar Diesel tractor, serial number 1C1, was built in San Leandro on August 1, 1931, and sold to a California farmer located between Salinas and King City. This farmer had agreed to purchase the tractor if he was satisfied with how it performed harvesting his beans. The farmer wasn't satisfied, so the deal fell through and 1C1 went back to the factory.[11] It was later sold to the Harms Brothers of Sacramento. The assembly of the second tractor, 1C2, followed. Tractor demonstrations started in September. W. C. Schuder of Woodland, California, attended a demonstration with his sons. On September 14, 1931, Schuder signed up for tractor 1C2 to use on his 2,000 acres in the Woodland area. Schuder was a longtime owner of Best Tracklayers and was confident that should problems arise, he would receive

7. Personal interview with Daniel G. Best (April 6, 2006).
8. "Changes at the Caterpillar Tractor Co.," press release. Caterpillar Tractor Co. (June 24, 1930).
9. "The Diesel Sixty," letter to Caterpillar dealers (August 31, 1931).
10. Ibid.
11. Milt Davies interview, F. Hal Higgins Collection, Special Collections, University of California Library, Davis.

the same fair treatment he had known from Best and later Caterpillar.[12]

The main production of the Diesel Sixty was at the East Peoria plant, but the "heart" of the engine, the fuel-injection pumps, was made at San Leandro. The first tractors were delivered with a Bosch fuel-injection system. While the Bosch system was serviceable, it did not totally meet the standards set by the company. Due to the minute tolerances required for optimal performance of the Caterpillar Diesel engine, it had been decided that all of the exacting work on pump and injector settings would be done at the factory, eliminating the need to do so in the field. Best and Starr stressed that this interchangeability had to occur. German engineers were brought to San Leandro to solve this problem. According to Oscar Starr, when the engineers were asked about how they worked out the interchangeable issue, they replied, "We don't." Starr said, "We do," and Caterpillar researchers were given the task to create a system to meet the high standards demanded.[13] Not only did a fuel-injection system have to meet those standards, but a way of mass producing it while maintaining those standards had to be found. Enter Byron Williford. Williford had been the superintendent of the machine shop back in the C.L. Best Tractor Co. days and continued to work for the Caterpillar Tractor Co. as the factory manager at the San Leandro plant. All of his knowledge and engineering skills were called upon to get the Caterpillar fuel-injection system into production. As before, if the production equipment didn't exist, Caterpillar engineers had to design and build it. The fuel-injector pumps, while controlled by Bosch but with manufacturing rights given to Caterpillar, required an incredibly fine tolerance of within twenty-five millionths of an inch and were required to hold their pressure of 1,800 pounds per square inch for a period of six minutes. This overengineering provided a safety margin for any possible emergencies.[14] The machining involved with achiev-

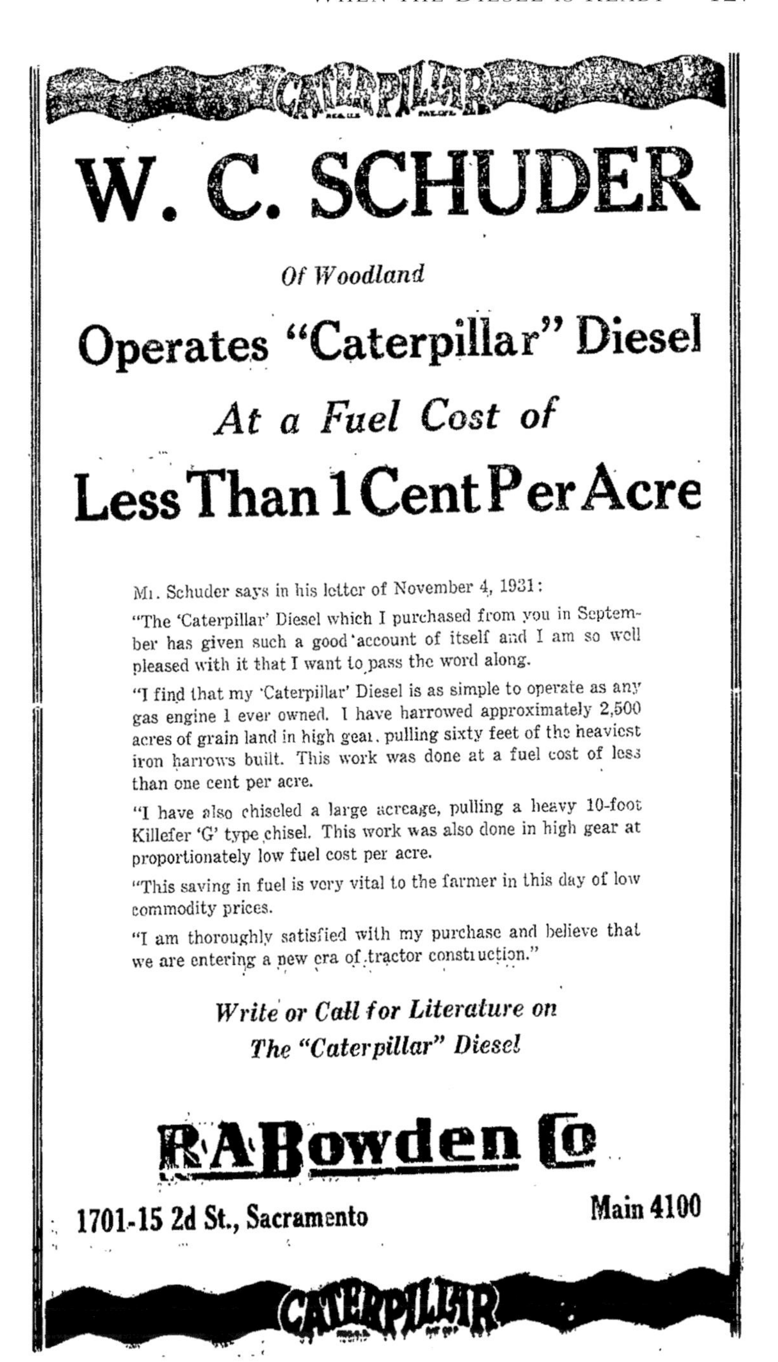

W.C. Schuder, a long-time Best customer, purchased Caterpillar Diesel tractor S/N 1C2 and his experiences were used in company advertisements. Woodland Daily Democrat, *November 4, 1931.*

12. "Owners of First Caterpillar Diesel Cut Fuel Costs," *Woodland Daily Democrat* (July 11, 1936): 3.

13. Oscar Starr interview, F. Hal Higgins Collection, Special Collections, University of California Library, Davis.

14. "Farm Costs Revolutionized by New Diesel Caterpillars," *Bakersfield Californian* (August 11, 1933): 15.

ing those close tolerances more closely resembled work done by lens grinders than machine shop workers. The operation was described in the May 1938 *Fortune* magazine article:

> *For this precision work, in which the tiniest particle of dust in the air is an industrial hazard, Caterpillar uses a shop in San Leandro, California. It looks more like a laboratory than a factory; its interior is painted white, and the employees wear clean, white aprons. Abrasive diamond dust is obtained for this work by mixing ordinary diamond dust with water, allowing the coarser particles to settle out, and then evaporating the water from the remainder; and the resulting powder is as soft as talc in the hand. Grinders turn at 60,000 revolutions per minute—so fast that the only way to measure the r.p.m. is to compare their shrill whine with the sound emitted by a high-frequency tuning fork.*[15]

Byron Williford again came through for his employer, just as he had when C. L. Best introduced the Best Sixty. The fuel-injection system that the Caterpillar Diesel engine required was in production by early 1932 and by tractor serial number 1C48, the true interchangeability C. L. Best and Oscar Starr insisted upon had been achieved.

Along with all engine refinements and changes in production, the Caterpillar Tractor Co. was still deeply affected by the continuation of the Great Depression. In 1932, the company posted a loss of $1.6 million. Economies in the company started at the highest levels. The year 1933 brought a salary cut of 20 percent to the top executives. The work schedule was trimmed to four days per week, and the total number of employees was reduced. But through this difficult time, the emphasis on research and development of new products never wavered. As a result of that concentration, in 1933, Caterpillar was able to offer buyers a product line that featured all new or redesigned equipment.[16] Even in the depths of the Great Depression, the fuel savings of the Diesel tractor when compared to a similarly sized gasoline-powered tractor was great enough to sell machines. Two major engine problems were yet to be solved: scoring of the cylinders and ineffectiveness of existing lubricating oils. One solution came from field observations, while the other involved major oil companies and the Caterpillar unrelenting demand for quality.

A farmer from Corcoran, California, had taken delivery of his first-generation Diesel tractor too late in the season for drawbar field work, so he used it for stationary work pumping water. He experienced no problems with the tractor. The other 1C tractors had encountered serious problems with scoring of the walls of the four cylinders and piston rings sticking when under heavy loads. When the San Leandro engineers did their take-down engine inspection, they noted that one of the precombustion chambers [burner tubes] had blown out the standard four holes and was operating with only a single square hole. They replaced the damaged one with the standard four-hole precombustion chamber. As the season progressed, the farmer used his Diesel tractor for plowing. Five hundred hours later, the San Leandro facility received notice that the tractor was down. Upon inspection, the precombustion chambers were intact, and all four cylinders had experienced the scoring found in the other Diesel Sixty tractors. As related by Oscar Starr in an interview with F. Hal Higgins:

> *So, as this liner [cylinder] scoring continued, I asked Peoria over the phone one day, why not make up a burner with but one hole. Finally, they did. No trouble. That solved it.*[17]

The final stumbling block was one that the Caterpillar Tractor Co. couldn't solve alone. In investigating and repairing the failures of different engine parts, the San Leandro engineers came to the conclusion that many of the failures were not related to design flaws, but to a lack of suitable lubricating oil. In 1933, the Caterpillar Tractor Co. introduced its next generation of Diesel-powered tractors. These tractors, the Diesel Thirty-Five, Diesel Fifty, and Diesel Seventy, all featured a 5¼" bore engine with an 8" stroke and ran at 850 RPMs. To obtain the required

15. "The Cat," *Fortune* 9, no. 5 (May 1938): 87.

16. "Depression-Era Sales Helped by Large Orders from USSR," *Caterpillar World: Century of Change* (May 1984): 29.

17. Oscar Starr interview, F. Hal Higgins Collection, Special Collections, University of California Library, Davis.

horsepower, additional cylinders were used: three, four, and six. With the increased horsepower and the increased motor speed, the lubrication problems worsened. Oils from all of the major companies were tried. All failed.

The Caterpillar engineers were looking at all possible solutions. They even went as far as taking advice from the CEO's son. Daniel G. Best, son of C. L. Best, was a farmer in Woodland, California, located in the Sacramento Valley. Dan was using a three-cylinder Caterpillar Diesel for his farm work:

> *I used that tractor all summer and didn't have any troubles. So I called Dad and told him that I didn't have any engine troubles and that maybe he should try the oil I used instead of some from the big companies. "By God," he said, "We'll look into it." And they did. See, the air is drier up here around Woodland, but when they tried in the more humid air around San Leandro, the troubles came right back.*[18]

The next step the company took was to invite all of the major oil companies to send their engineers and chemists to the Research Division in San Leandro to show them just how poorly their lubricating oils withstood the rigors of the Diesel engine. The men who arrived were given total access to the Caterpillar lab. The search for the solution turned into an unofficial competition. After months of experimenting, the noncompetition was over. A group of oil additives was found that appeared to be promising. The new oils were known as compounded lubricating oils. This new type of oils did indeed solve the final major problem associated with the Caterpillar Diesel engines.[19] The compounded oils became the standard for the industry thanks to Caterpillar's unrelenting demand for excellence.

With the lubrication problem solved, the Caterpillar Tractor Co. focused on the future. Their Diesel engines, whether used in tractors, road machinery, or as power units, became the industry standard. The first Diesel engine was announced on August 31, 1931. It took until September 23, 1933, to produce the first one

Get the Proof—
Demand a
SHOW-DOWN

The "Caterpillar" DIESEL is making new records of low-cost operation wherever it goes, winning new owners constantly by its performance record. In California 1220 "Caterpillar" DIESELS are used in agriculture, industry, contracting and lumbering because the "Caterpillar" DIESEL is the leading source of sure, dependable power at rock-bottom costs.

The following owners in our territory asked for the proof. They got it . . . and as a result bought "Caterpillar" DIESELS.

YOU, TOO, ARE ENTITLED TO A SHOW-DOWN—DEMAND IT.

HERE THEY ARE—
53 Owners of 81
Caterpillar
DIESELS

The fleet owners in this list first bought one "Caterpillar" DIESEL. So well did it perform that they added one after another until one fleet owner now has eleven "Caterpillar" DIESELS.

Owner	No.	Town
Amador County	(3)	Volcano
Carden Bros.	(1)	Davis
Capital Dredging Co.	(1)	Folsom
C. A. Baker	(1)	N. Sacramento
C. C. & C. L. Bennett	(1)	Sacramento
Dan Best	(2)	Woodland
E. Burquist	(1)	Knights L'd'g.
R. L. Button	(2)	Winters
E. L. Dexter	(1)	Winters
Durst Bros.	(2)	Capay
El Dorado Ranch	(1)	Knights L'd'g.
Fisher & Rich	(1)	Sacramento
F Frates	(1)	Sacramento
A. Fuschlin	(1)	Knights L'd'g.
M. R. Giguiere	(1)	Yolo
J. Harlan	(1)	Woodland
Harms Bros.	(4)	Sacramento
C. W. Hayenga	(1)	Galt
D. & L. Hayes	(1)	Woodland
Hershey Estate	(2)	Woodland
Johnston Bros.	(1)	Galt
H. G. Kaupke	(1)	Woodland
F. E. King	(1)	Sacramento
B Krasnow	(1)	Sacramento
Lord & Bishop	(1)	Sacramento
J. Mahon & Son	(1)	Elk Grove
J. G. Mast	(1)	Woodland
Mrs. V. Megonigal	(1)	Arbuckle
L. M. Miller	(1)	Grimes
B. McCormack	(1)	Rio Vista
D. McDonald	(11)	Sacramento
T. McEnerney	(1)	Galt
Natomas Company	(1)	Sacramento
C. Parella	(1)	Sacramento
Geo. Pollock Co.	(4)	Sacramento
Rogers Bros. Seed Co.	(1)	W. Sacramento
Theo. H. Roth	(1)	Woodland
Rominger Bros.	(1)	Woodland
W. C. Schuder	(1)	Woodland
T. B. Sills	(1)	Rio Linda
Sims & Sturges	(1)	Elk Grove
J. W. Small	(1)	Galt
W. E. Stewart	(1)	Rio Vista
T. E. Tadlock	(1)	Winters
A. Tiechert & Son	(5)	Sacramento
U. S. Engineers	(1)	Sacramento
L Ulrich	(1)	Woodland
Carl Vanderford	(1)	Rio Linda
N. Wallace	(1)	Woodland
E. L. Wallace	(2)	Woodland
W. C. Watson	(2)	Ione
West. Land & Rec. Co.	(1)	Knights L'd'g.
Yolo County	(1)	Woodland

WEAVER-RYE
TRACTOR CO.
TRACTORS CATERPILLAR EQUIPMENT
21 Main Street, Woodland Phone 830
Agents, Oliver, Killefer, and Dyrr Tractor Tools

10,000
"Caterpillar" Diesel
Engines Can't Be Wrong

The completion of "Caterpillar" Diesel No. 10,000 on Nov. 13th was more than a ceremony. It meant that in a little more than four years, Caterpillar Tractor Co., first American manufacturer of a Diesel Tractor, had established a new all-time record for Diesel engine production.

Note the speed of this growth:—

First Diesel announced Aug. 31, 1931
1,000th Diesel built Sept. 23, 1933
2,000th Diesel built Jan. 16, 1934
3,000th Diesel built April 4, 1934
4,000th Diesel built Aug. 1, 1934
5,000th Diesel built Nov. 5, 1934
6,000th Diesel built Feb. 4, 1935
7,000th Diesel built May 6, 1935
8,000th Diesel built July 11, 1935
9,000th Diesel built Sept. 16, 1935
10,000th Diesel built Nov. 13, 1935

In 1934 "Caterpillar"-produced almost one-third of the 750,000 Diesel horsepower manufactured in the United States. "Caterpillar" Diesel production for nearly 11 months in 1935 shows a 30% gain over its 1934 figure. Power users everywhere want the "Caterpillar" Diesel because of its low maintenance cost and its ability to burn low cost fuel oil and less of it.

Demand a SHOWDOWN Before You Buy a Tractor.

WEAVER-RYE
TRACTOR CO.
TRACTORS CATERPILLAR EQUIPMENT
21 Main St. Woodland Phone 830
John Deere Tractors and Implements
Killefer and Dyrr Tractor Tools

(Above left) *Weaver-Rye dealership advertisement for the Caterpillar Diesel listing various owners. Note that Dan Best, son of C. L. Best, is listed as owning two tractors.* Woodland Daily Democrat, *September 7, 1935.* (Above right) *Newspaper advertisement showing the timeline to the completion of the 10,000th Caterpillar Diesel engine.* Woodland Daily Democrat, *December 28, 1935.*

18. Personal interview with Daniel G. Best (April 6, 2006).

19. E. E. Wickersham, unpublished manuscript (1940), Special Collections, University of California Library, Davis.

thousand. Less than fourteen months later, November 5, 1934, saw the five thousandth engine completed. Three hundred seventy-three days later, the production doubled with the ten thousandth Caterpillar Diesel engine completed at the Peoria plant on November 13, 1935. By the time that ten thousandth Diesel-powered tractor rolled off the production line, the Caterpillar Tractor Co. had produced a total of 640,450 Diesel horsepower. Their Diesel-powered equipment was on the job in seventy-two countries across the globe, but quest for improved models didn't wane. In later 1935, the company introduced a new line of Diesel tractors: the RD8, RD7, RD6, and later the RD4.[20]

The trials of the Great Depression were past; the Caterpillar Diesel was the acknowledged leader in its field; the company was growing in size and profitability; and C. L. Best had seen his vision for the company he helped create fulfilled. By standing firm in his convictions of "simpler, more efficient, longer life," Best set the standards that would guide the Caterpillar Tractor Co. far into the future.[21] By the mid-1930s, Clarence Leo Best, inventor, entrepreneur, and CEO, removed himself from the day-to-day operations of the company. However, he remained keenly interested and offered prudent counsel when needed. But there was much more to Best than just his mechanical aptitude and business sense. Oregon born, but California raised, Leo Best, the man, also had quite a story to tell.

Caterpillar Diesel tractors serial numbers 1C1 through 1C157, where first known as the Diesel Sixty then the Standard Diesel tractor. Several years after production ceased, they were referred to as the Diesel Sixty-five tractor. Author's Collection.

20. "The 10,000th Caterpillar Diesel Rolls Off the Assembly Line," *Fresno Bee* (December 8, 1935): 8.

21. "A Tribute to Clarence Leo Best," *Caterpillar News Service* (September 22, 1951): 3.

Leo Best

When you are born into a family where the patriarch is a mechanical genius, two things can occur: You can stay safely in his shadow and be content, or you can take a chance and rise above his accomplishments. Leo Best did not stay safely in his father's shadow. Daniel Best realized early on that his son Leo was mechanically gifted. He exposed Leo to every aspect of his business: from concept and design to pattern making and foundry work; from the assembly floor to sales and delivery. Not only could Leo take a concept and visualize the steps required to the finished product, but he had an innate business sense that allowed him to make the right decisions at the right times. Oscar Starr, first an employee and later a brother-in-law, neighbor, and good friend, commented that while Leo was able to do it all, his true strength lay in his vision of the future, or the big picture.

Although designing and manufacturing machines consumed vast quantities of his time, Leo had many other outside interests. As a young man growing up in San Leandro, Leo was proud of his connection to the city. He was civic minded and early on played an active role in the community. In 1897, the Amateur Orchestra of San Leandro was organized with Leo as its librarian. He played the clarinet with the orchestra. A few years later, the editor of the San Leandro paper submitted "A List of Eligible Young Men for the Girls to Choose From." Leo Best was on the list. He served as fire chief in 1900 and later was on the city council, the library board, and his church council. Leo was competitive and enjoyed playing baseball. He helped organize the first ball club in San Leandro and played short stop. Later his company would sponsor the semiprofessional team, Best Tractors. Competition between the Best Tractors and the Holt Caterpillars was spirited. Fans from both San Leandro and Stockton supported their teams by caravanning between the cities to the ball games. The year 1919 saw the Best factory getting involved. To show their support for the Tractors, the crew made a big anvil, which when struck by team supporters could be heard for blocks.

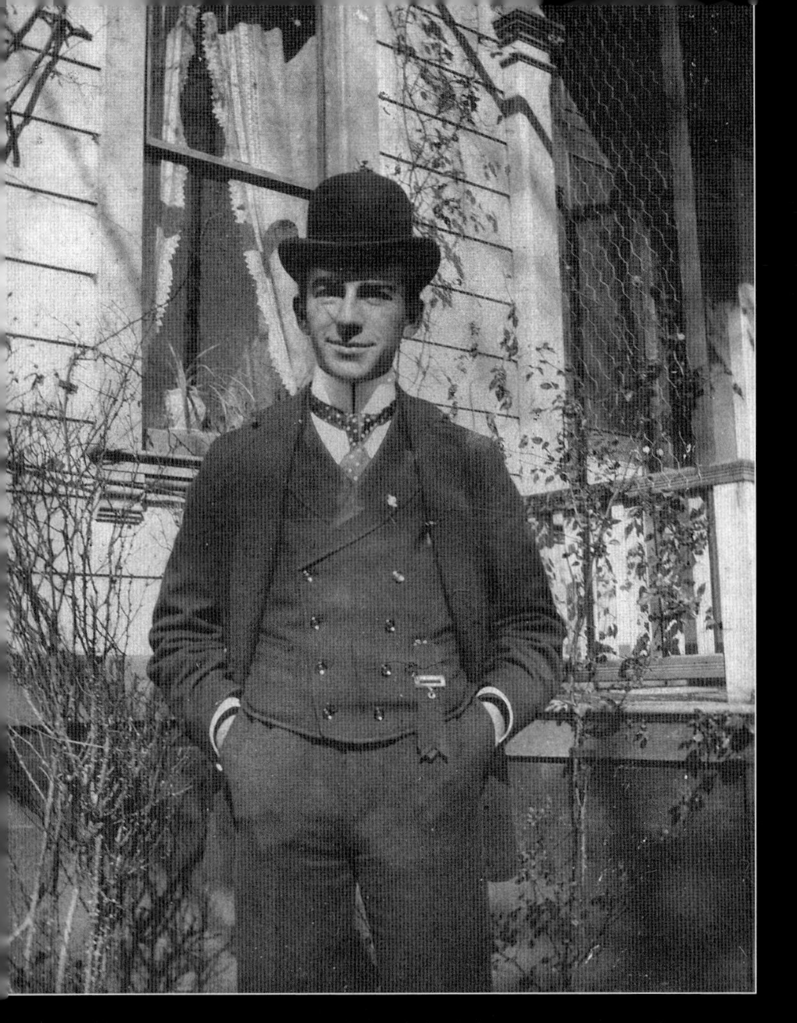

Above left) *Leo Best standing outside his father's house in San Leandro, California.* 1895. Photo courtesy of Dan G. Best II. (Above right) *The C.L. Best Tractors baseball team holding the Mission League championship pennant.* 1918. San Leandro Public Library Historical Photograph Collection #00481. Lower right) *The C. L. Best family. C. L. is seated; his wife Pearl is standing to his left with their daughter Elizabeth (Betty). Young Dan is seated on his father's lap, looking straight at the camera.* 1913. Photo courtesy of Dan G. Best II.

On the personal side, Clarence Leo Best married Pearl Margaret Gray on December 11, 1900, in a ceremony held at his father's house in San Leandro. The popular young couple played host to numerous social events in the area. Leo continued to work at the Best Manufacturing Company, and life was good. In 1907, they welcomed their first child, Elizabeth (Betty), into their family. But in 1908, things began to change. Daniel sold his company to the Holt Manufacturing Company. Leo continued with the company, confident that the knowledge and skills he would bring and the contributions he would make would be valued by the new owners. By January of 1910, Leo was frustrated that his authority was being usurped and felt relegated to being just a figurehead in the company. He tendered his resignation on January 15, and by March 27, the building permit for the new C.L. Best Gas Traction Co. factory in Elmhurst had been issued.

Starting the new business was all consuming. Financial backers had to be found and convinced by Best that he had a viable product and was capable of getting that product on the market. He had to get a manufacturing facility constructed and equipped for production, and he had to design the product. Finances were tight. Stories of selling tractors to make payroll were not exaggerations. But Leo and his fledgling company persevered and expanded in spite of the hardships.

Leo and Pearl's son, Daniel Gray Best, was born in November 1912. The family was now living on a thirty-acre tract of land in San Leandro known as the A. C. Peachey property, which included the Peralta house. This property was only a few blocks from the elder Daniel's house on West Estudillo. In the later years, after Leo moved his company back to San Leandro, his son Dan remembered that "he would leave the house in the morning, walk the block to the railroad tracks and follow those tracks to the plant on Davis Street so he could get some exercise." The thirty acres were owned by the "Old Man," as the elder Daniel Best was known to his grandson. "We didn't mean anything bad by the name," said Dan, "it was just what we called him." On some other land owned by the Old Man were some of the cherry trees for which San Leandro was famous. Those trees and their ripened cherries led to a teaching moment for the younger Daniel. As told by Dan G.:

We were living at the Peachey Place in San Leandro. There must have been twenty acres of cherry orchards that the Old Man owned. So Dad and the Old Man decided to teach me a lesson in economics and have me sell cherries. Even though we owned the trees, Dad made me go to the Old Man and buy the cherries. Grandfather was so hard of hearing that you had to shout at him and then he shouted back. I hated to shout at him. He kind of scared me. I got the cherries, bagged them up, and set up a little stand. I used two packing crates with a board across for my stand. I made a sign that said, "Cherries—.25." I did good selling cherries. I had to go buy more from the Old Man. After a while, a little Portuguese girl set up a stand on the same street but before mine. All the people stopped at her stand because it was first. So, I made a new sign that said, "Cherries—.15." You could see people starting to slow down for her stand, then notice that my cherries cost less. After about three or four days, she gave up. I changed the sign back to "Cherries—.25."[1]

The "Peachey Place" era for the Best family held good and bad memories. Leo was driven to see his business survive. According to his son, Leo was hard pressed to turn off thoughts of tractors:

I remember packing into the Sierras to go hunting. I'd wake up at night and look over to where Dad was supposed to be sleeping. You could see the red tip of his cigarette glowing as he was sitting up thinking. He was always thinking. Even in the house, he would be staring into space, and I'd know to leave him alone because he was working out something in his mind.[2]

The three generations of Best men spent time together hunting and fishing in the Sierras and also at their duck camp near Colusa. The Old Man, while now hard of hearing, had lost noth-

1. Personal interview with Daniel G. Best (April 6, 2006).

2. Ibid.

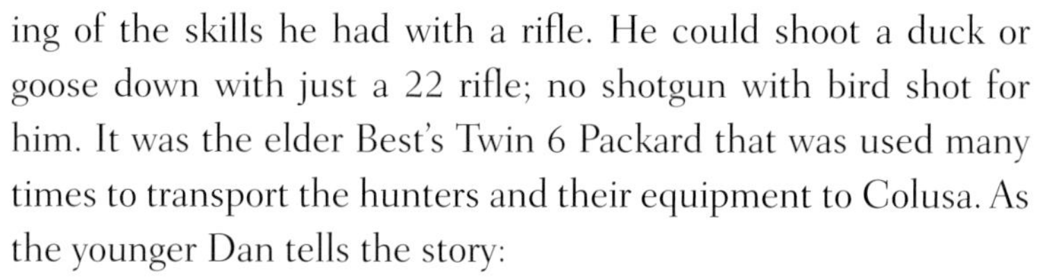

ing of the skills he had with a rifle. He could shoot a duck or goose down with just a 22 rifle; no shotgun with bird shot for him. It was the elder Best's Twin 6 Packard that was used many times to transport the hunters and their equipment to Colusa. As the younger Dan tells the story:

We were going goose hunting up at Colusa. The Old man was driving his Packard and the storm curtains were on. The curtains had an armhole for signaling your turns. I was sitting in the back seat, behind the Old Man. Signaling wasn't the only thing he used that armhole for. He was chewing tobacco and spitting out the curtain hole. By the time we got to Colusa, I couldn't see out anymore.

> *I must have been 11 years old or so [about 1923]. We would always have a big feed before we left duck camp to come home. Chili, I think. Well, everyone really ate. Dad would be driving the old Packard home from Colusa and pretty soon I'd hear, "Dan, are you awake?" Since I was the only one who wasn't sleepy, he'd switch places with me and let me drive. Imagine that, a little kid driving that big*

(Above left) *The Peachey Place as it looks today. In 1926, the house was purchased by the Alta Mira Club, a social and cultural society in San Leandro.* Author's collection. (Above right) *This photo was taken by C.L. Best of a relaxed moment at a deer camp in the Sierras. Daniel Best is seated second from right.* San Leandro Public Library Historical Photograph Collection #01618. (Lower right) *Daniel Best was proud of his Packard automobiles. In a letter dated April 24, 1915, written to his brother Benjamin in St. Francisville, Missouri, Daniel invited Benjamin to California to visit and take in the "great Panama Exposition. I have a large automobile and can show you over the country. I sold my business out a few years ago and am now at leisure so we can put in the time together."* Photo and letter courtesy of Frank Best.

Packard. I'd start driving; he'd make sure it was shifted into high, and then I was on my own.[3]

Leo and his son shared a love of hunting. The passion shared by Leo and his daughter, Betty, was horses. Leo was pleased to provide Betty with excellent horseflesh. She was known as a world-class horsewoman and won many awards for her horses and her showmanship. One year, Betty and her horse Congo were chosen to lead the San Leandro Cherry Festival parade. It was in an earlier parade that Grandfather Daniel proudly drove one of his steam-traction engines with the canopy bedecked with flowers. While Betty and Leo enjoyed the long horseback trips packing into the Sierras, Dan tolerated the horses. They were a means of transportation to him. According to Dan:

Dad really loved horses. I remember he kept some on his ranch at Mission San Jose. One of them was so old he didn't have any teeth left to chew oats. So Dad had oatmeal cooked and fed it to him. Imagine, feeding an old horse mush.[4]

The Peachey Place saw some sibling squabbles that Leo had unwittingly started. As Dan told the story:

One time, Dad made me a dynamo with 4 spark plugs and a handle you could turn. I could watch the spark plugs fire. That set me to thinking. So, I wired it to the metal door mat and the brass door knob and attached a long wire. I hid and when my sister got a hold of the knob, I cranked the machine. It's a good thing I could run faster than she could. I could always run to the orchard and climb up a tree. I would go high enough up so that she wouldn't follow. Another time, I wove a thin wire into the fabric cushion of the dining room chair. I waited until she sat down, excused myself for some reason, and gave the dynamo a good crank.

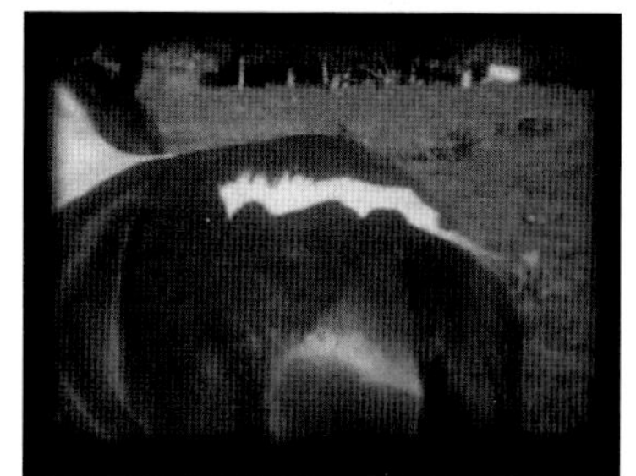

(Clockwise from top left) Leo examining the hoof of his daughter Betty's horse, Red. Billy the Caterpillar horse. Note the wavy line on horse's rump. Close up of the white marking on Billy. Leo taking his turn at camp cooking. Leo after a successful hunting trip. Leo and friend at Yosemite Park in California.

She really jumped. But Dad couldn't say much. He gave me the thing in the first place.[5]

While Leo the father was making memories with his two children, Leo the inventor and businessman had his hands full. February 1915 saw the Holt Manufacturing Company bring a

3. Ibid.

4. Ibid.

5. Ibid.

The crane of the C.L. Best Gas Traction Co. that spanned Davis Street in San Leandro, California. The automobile is Daniel Best's Packard. Daniel Best is standing in the middle of the photo in the dark three-piece suit wearing the broad-brimmed hat. C. L. Best is standing to Daniel's left. The caption in the upper right of the photo reads "San Leandro-1916. C.L.B." Special Collections, University of California Library, Davis.

patent infringement lawsuit against Best. This suit would continue for nearly four years. By 1916, the C.L. Best Gas Traction Co. was out of room to expand in Elmhurst and actively looking for a new site. That new site was in San Leandro, on the same property where Daniel had established his company in 1885. This put Leo living only blocks from the factory and allowed him that quiet walk to work. The year 1917 brought the loss of control of the company to C. A. Hawkins. During this time, Leo was actively involved with the research and development of the Model Sixty tractor. Gaining back control of the company he started, working out the design of the Sixty, and dealing with the protracted and expensive legal battle with Holt left little free time. This took a heavy toll on Leo and Pearl's marriage, leading them to consider a divorce. When young Dan Best expressed his dismay at that possibility, the couple stayed together.

The situation at home remained strained, but times improved for Leo and his business. Hawkins was ousted and control went back to Best. The Holt/Best lawsuit was settled in Holt's favor,

but Best came out with $200,000 cash as a payment for the Best patent rights transferred to Holt control. This cash infusion put the Best concern on firm financial footing, and when the revolutionary Best Model Sixty tractor was introduced in 1918, the future Leo Best had envisioned arrived. Leo was able to share these triumphs with his mentor and his friend: his father, Daniel Best. Imagine the conversations held between father and son during those dark times. What counsel did the elder Best, who himself had failed at numerous business ventures before succeeding, give his son? Think of the discussions when the topic was tractors. Imagine the smile on Daniel's face when he sat in

(Above left) *Best Sixty Tracklayer S/N 101A. This tractor, owned by Tom Madden of Paso Robles, California, was restored to its original June 1919 condition by Ed Claessen of Waverly, Minnesota. 2007.* Author's photo. (Above right) *Daniel Best in a photo taken near the end of his 85 years.* Photo courtesy of Frank Best. (Lower left) *"The Mansion" as C. L. Best called the estate he had built in the hills of Piedmont, California.* Photo courtesy of Dan G. Best.

Sixty tractor removing almond trees on the C. L. Best property in San Leandro, California. Note C. L. Best on the right side of the photo standing near the pole. This tractor, with a used radiator core, was taken for a drive by Dan G. Best and his friend. Special Collections, University of California Library, Davis.

The shop located on the ranch owned by C. L. and Dan G. Best. It included an overhead crane from the original Best Manufacturing Co. and machine shop tools salvaged from the Holt Manufacturing Company's Stockton plant. This shop is still in active service. Photo courtesy of Dan G. Best.

the operator's seat and drove the newly assembled prototype Best Sixty No. 101A for the first time. You can almost hear him saying, "You did good, son. This is the one. You've got it right." Leo lost his father on August 23, 1923, after a short illness.

With the Sixty and the Thirty tractors now in production, the financial burdens eased. Leo was still closely involved with the San Leandro plant, but demands for his time lessened. His wife, Pearl, was considered a society matron, known for her lavish entertaining skills. With the incorporation of the Caterpillar Tractor Co. and Leo becoming chairman of the board, the Peralta house on the Peachey property no longer fulfilled Pearl's ideas of how and where a tractor magnate should live. In 1925, Leo, hoping to appease Pearl, bought eight acres of land in Upper Piedmont and construction began on the new Best estate. The house would have 8,492 square feet of living space and be Italian Renaissance in design. The main dwelling would be built on the hilltop with a stable for some of Leo and Betty's favorite horses included on the lower acreage.

Leo owned the house on the Peachey property, but the orchard surrounding the house went to the entire Best family upon Daniel's death in 1923. With the City of San Leandro growing, the land occupied by the orchard was more valuable as building sites than as an active orchard. Leo arranged for the power used to pull out the trees. As Daniel G. Best remembered:

> *Dad had a Sixty brought up from the factory. It wasn't that far. I remember it had street plates on the tracks. Dad had them put in an old radiator core and they used an old fuel tank. I remember that the tractor wasn't all painted yet. When the work was done, they took it back to the factory and changed the radiator and fuel tank and finished painting it. You don't suppose that some poor bastard paid new price for a used machine?*
>
> *While that Sixty was working at the Peachey Place, Dad told the operator to put me up there and show me how to drive it. My friend and I saw how he started the tractor. So later, when no one was around, we decided we would start it and drive it around. I had to have both hands on the steering clutch lever and brace my foot on the column to pull the lever. We drove it all around but made sure to park it back exactly where it had been. Of course, we didn't think about all the tracks we had made.*[6]

6. Ibid.

Leo moved his family to Piedmont while maintaining his office at the Caterpillar facility in San Leandro. It was during this time that the research on Diesel power was increased. Leo was keenly involved with the research and testing as he was sure that the Diesel engine would be the power for the future of the Caterpillar Tractor Company. For Pearl, this dedication to the company seemed a repeat of years past. By 1929, the house on the hill in Piedmont was offered for sale, and Pearl moved to the Woodland, California, area. The couple had grown apart over the years. In August 1930, the divorce case of Clarence Leo Best and Pearl Margaret Gray Best was heard in superior court in Woodland in a secret hearing. The court granted Pearl an interlocutory decree of divorce. As part of the settlement, Leo purchased the former C. Q. Nelson estate and had the house modernized, a swimming pool installed, and various outbuildings erected. Pearl, daughter Betty, and son Dan, who was still in high school, moved into the estate located on the Knights Landing highway.

The next year, Dan was enrolled in college at Davis, California, and lasted one semester before realizing it wasn't for him. He remembered his father confronting him and asking about his plans: "He made sure I knew he expected me to do something now that I wasn't going to college anymore. I told him I wanted to farm." A partnership was formed between father and son, and Daniel G. Best was a farmer. This partnership allowed an interesting parade of tractors to arrive in Woodland over the next ten to fifteen years. According to Dan:

> *When I was going to be doing something at the ranch that needed different power, I would mention it to Dad and his response was, "I'll look around." That's how I got the Diesel Seventy. It was an experimental that the plant was done with. That one got scrapped when I was done with it. I had a Diesel Thirty-Five. It was experimental, too. I had to keep track of fuel consumption and other fluids that were used on that one.*[7]

As the years went by, other unique tractors appeared at the ranch. Dan had a 5E Diesel Forty he used and an experimental Twenty that had been on Leo's ranch at Livingston. When the Twenty arrived, it was accompanied by a Sixty that had also seen use on Leo's ranch. Dan had an experimental six-cylinder D6, and later, coming from the Caterpillar Proving Ground at Phoenix, Arizona, two No. 70 scrapers and a D7 (all experimental). A modern shop building with a machine shop was constructed. When talking about the Diesel tractors that were used on the ranch, Dan shared an incident he witnessed in the late 1920s:

> *I was in Dad's office one day when Henry Kaiser came in. I always felt Henry was a forceful promoter. He had an Atlas Imperial Diesel motor that he felt Dad should use in his tractors. Henry proceeded to tell Dad why Caterpillar should use his motor instead of continuing to develop one of their own. Dad told Henry, "Thank you, but I'm not interested," and that he was busy and sent Henry out the door. Henry bought some Sixties and put his motors in them. That didn't work out at all.*[8]

The father–son partnership continued until 1948 when Leo finally said to Dan, "It's obvious I'm not going to be farming, so you might as well buy out my half." And Dan did. In the appraisal of the ranch prepared in July 1948, the appraiser commented that:

> *The . . . ranch is well equipped with large quantities of modern, heavy, mechanized farming equipment. Due to the relationship of the owners with the Caterpillar Tractor Company, the ranch has been used as a testing ground for heavy earth moving equipment as well as for design changes in the Caterpillar Tractor. The conditions under which the experimental equipment are used specifically prevent resale and requires final junking upon completion of the experiment. Therefore, experimental equipment is not listed herein and no value has been given to such*

7. Ibid.

8. Ibid.

This replica of the original C. L. Best nugget collection is on display in Downieville, California. The original nuggets from the Ruby Mine are on display at the Los Angeles County Museum of Natural History. Author's photo.

> *equipment. A very extensive shop building with a complete machine shop is maintained to service the farm equipment as well as the experimental pieces.*[9]

On August 29, 1931, Leo Best married Irene Wagness Hansen in San Jose, California. Irene and Hazel Starr were sisters. Hazel was the wife of Oscar Starr, longtime business associate, neighbor, and good friend to Leo. While Leo stayed keenly interested in tractor design and development, he had other interests, too. He owned three cattle ranches in Merced and Mariposa Counties; at his ranch at Livingston, California, he raised show horses, cattle, and hogs. He was an avid hunter, with a membership in a duck club at Colusa, California, and he also enjoyed packing into the Sierras on horses for hunting. He conducted oil explorations in California and Nevada. Leo also leased and owned various gold mines in California. Most notable was the Ruby Mine near Downieville. It was 1934 when he reopened the Ruby. The mine was in active production from 1937 to 1942. While it was not a large strike, it did produce an amazing amount of specimen nuggets. One gold nugget weighed 3¼ pounds and was the size of a softball. Leo saved the best of the best for his collection: 159 gold nuggets weighing nearly one thousand ounces and assayed as 96 percent pure. In 1942, during World War II, the U.S. government froze gold prices. The Ruby mine was shut down because it was no longer profitable. Dan Best recalled the Caterpillar tractor used at the mine:

> *Dad had a Diesel Seventy-Five he used to improve and maintain the road to the mine. The ore had to be brought to town to the stamping mill. When he closed down the mine, he sent the tractor to me in Woodland.*
>
> *At that time, the hospital here in Woodland had a six-cylinder 5¾ inch bore Cat generator set that was being swapped out. Dad told me to get that engine, put it in the Seventy-Five, and then I'd have a D8. I did, and used it on the ranch.*[10]

After living on his ranch at Mission San Jose, near Oakland, for twenty-three years, Leo and Irene moved to their new home in Walnut Creek. He made sure he had space set aside for one of his most important possessions: his drafting table. It was there that Leo continued to put his never-ending stream of ideas on paper. Ever the visionary, he even predicted that sometime in the future tractors would be powered by the sun—amazing for a man who started designing equipment when horses still provided much of the power. But not all of his drawings had to do with tractors. As remembered by Leo's grandson, Daniel G. Best II:

> *One Christmas, when Grandfather and Irene were visiting with us in Woodland, my sister Brenda said, "Grandfather, invent me something. I want to be a millionaire." So he took a napkin and did a sketch of a duck decoy with a retractable anchor line. Just like that: in no time at all.*[11]

9. Appraisal C. L. Best and Daniel G. Best ranch, Rush Farm Management Service (July 30, 1948), Daniel Best II collection.

10. Personal interview with Daniel G. Best (April 6, 2006).

11. Personal interview with Dan G. Best II (June 30, 2008).

This sketch is from the drafting table of Leo Best, January 20, 1937. The engine is located in the rear, a design not used by Caterpillar until the mid 1980s. Leo drew "a flashy grid to cover toolbox and focus attention on the front end" and a mirror to see the pull bar. The lettering on the front says Cat Diesel with the wavy red line. Caterpillar, Inc. archives. Courtesy of Dan G. Best II.

Dan II recalls other times when Leo and Irene came to visit:

> *I remember Grandfather and Irene coming to Woodland. They were always driven by their chauffeur. They traveled with their dogs—they were collies. I remember the adults were in the house, but I would go outside and play with those dogs.*[12]

Although Leo hadn't been at the San Leandro facility full time for many years, he was still well known and well liked in San Leandro. He had an easy-going manner. His son, Dan, recalled:

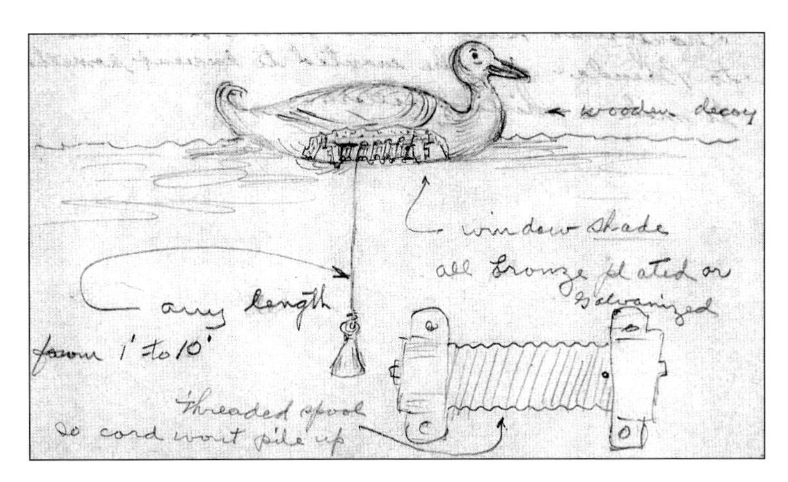

Duck decoy drawing sketched by Leo Best for his granddaughter Brenda Best. Sketch courtesy of John Weaver.

> *Dad had the knack of making friends everywhere he went. There were lots of times we'd be driving somewhere and we'd stop for lunch. I'd go to the restroom to wash my hands and by the time I got back, he'd be visiting with everyone in the place like he had known them for years. That was just the way he was.*[13]

San Leandro factory workers remembered stopping for a haircut at the neighborhood barbershop only to find the head of the company waiting for his turn to have his hair trimmed. Before long, everyone was talking and laughing. How you earned your living and what you were paid made no difference to Leo Best.

In September 1951, Leo and Dan were planning their deer hunting trip to the wilderness of Modoc County. Fall deer hunting was a time that father and son had enjoyed together for many years. The story continues in Dan's words:

> *Dad hadn't been feeling just right but didn't want to put off hunting. I told him he needed to go in and find out what was wrong. We could always go hunting next year. He said, OK, he'd go in and get checked out, but he wanted me to go hunting. So I went hunting and he went into the hospital in San Francisco. It was Saturday a little after breakfast. I heard a plane, and then saw it. I knew it wasn't anything*

12. Ibid.

13. Personal interview with Daniel G. Best (April 6, 2006).

Model of a Caterpillar Sixty tractor built by Pliny E. Holt in 1928 for the International Exposition in Seville, Spain. This 10" high x 16" long x 9" wide nickel-plated model is built on a 1/10 scale. Holt, with the help of other Caterpillar employees, built the model in his basement in about 2 ½ months. Original model is on display at the Haggin Museum, Stockton, California. Photo courtesy of the Haggin Museum, Stockton, California.

> *good. The pilot found us, circled, and then dropped a rock. A piece of paper tied to the rock. It said Dad was dead. We had to break camp, saddle the horses, and head back. When I got back, they told me that on the night Dad died, they went in and wanted to do some more tests. He told them that he was tired and wanted to rest, and that they could do the tests the next morning when he had rested and was fresh. A nurse went in a while later, and he was already gone. It was an aneurism in his heart. I think it was already going, that's why he was so tired.*[14]

Clarence Leo Best, inventor, businessman, tractor builder, husband, and father was dead at the age of seventy-three. In respect for the man instrumental to the founding and success of the Caterpillar Tractor Co., the San Leandro plant, with its one thousand workers, closed the entire day of his funeral service.

50-year pin presented to C. L. Best by the Caterpillar Tractor Co. Best was the only recipient of a 50 year. Courtesy Dan G. Best II.

Condolences and tributes were received from far and wide. The most evocative was the official tribute from the Caterpillar Tractor Co.:

> *The death of Clarence Leo Best, Chairman of the Board of Directors and member of the Executive Committee, has brought profound sorrow and a sense of great personal loss to his many friends and close associates.... He will long be remembered for his many contributions to the success of Caterpillar. Results of his practical engineering genius abound in the many products of the Company, in particular the track-type tractors.*
>
> *He [Leo] was instrumental in leading the Company into many years of engineering development that resulted in the adoption of the Diesel engine in our machines.... For him, it was not enough to have vision, but he had the persistence to forge ahead and shape the future.*

14. Ibid.

Many of us will remember him seated at his desk, feet up and hat cocked a jaunty angle. There he sat thinking in terms of improved machinery [and] searching for better ways. When he referred to the old Best Thirty and Sixty, his warm smile carried special meaning, for it was an accepted fact that these tractors contributed mightily to lay the foundation for the growth of the Company.... Leo Best had a good sense of humor.... He was genuinely modest and of a retiring nature, not one to speak of his accomplishments.

President Neumiller said, "In these words—'simpler, more efficient, longer life'—Leo Best leaves a rich heritage to all of us, an objective, a goal toward which to strive, as we should always remember these expressive words as we commemorate the passing of the only holder of the Caterpillar 50-year pin, a wise counselor and a very human friend, Leo Best."[15]

Leo Best. Photo taken at the Ruby Mine. Photo courtesy of Dan G. Best.

And so it ended: seventy-three years of life, fifty-four years in tractor manufacturing, and twenty-six years as chairman of the Board. From the beginnings of mechanized agriculture with horse-drawn combined harvesters to the refined power of a D8 Diesel tractor, from a fledgling company that maintained its existence with courage and determination to a company known and respected around the world, from Oregon to Oakland, with stops in Elmhurst, San Leandro, Mission San Jose, and Walnut Creek, C. L. Best met life's challenges with tenacity, fortitude, and a quiet self-assurance that with perseverance and hard work those challenges would be transformed into triumphs. The fundamental technology and engineering he pioneered may be overshadowed by technological advancements made by modern-day Caterpillar Inc., but how high could a skyscraper soar without solid footings below ground? C. L. Best provided the footings for the company that began in 1925. He would be both proud and humbled to see how it has soared.

15. "A Tribute to Clarence Leo Best," *Caterpillar News Service* (September 22, 1951): 1–3.

APPENDIX I

TRACTOR EVOLUTION

C. L. Best launched his Model A Sixty Track-Type Tractor in 1919. This tractor would become his signature machine and along with the smaller Model S Thirty Track-Type Tractor became the production foundation of the Caterpillar Tractor Co. in 1925. Always building from previous models and realizing the importance of parts commonality, Best and his engineers created a legacy that directed design and production of future track-type tractor models.

At the beginning of the twentieth century, tractor equipment manufacturers established claims each would make regarding horsepower and other attributes of their tractors. The product consumer had no accurate method to compare various manufactured makes and models. An impartial testing program was needed to accurately verify manufacturer claims. In 1919 the Nebraska State legislature passed the Tractor Test Bill. This bill was to "provide for official tests of gas, gasoline, kerosene, distillate or other liquid fuel traction engines in the State of Nebraska."[1] Any manufactured model tractor offered for sale in Nebraska would be subject to these tests, and the results would be available to the general public. The Nebraska Test lab was established at the University of Nebraska, and in 1920, Nebraska Tractor Tests were started. These tests soon established standards that were used throughout the nation.

To help readers not familiar with engine design, certain key references were used by manufacturers to denote the size and capabilities of an engine. The diameter of the cylinder (piston diameter) was known as the bore of the engine. Stroke referred to the distance the piston traveled on its way in the cylinder when the engine was running, and RPM referred to revolutions per minute of the crankshaft.

The following information traces the evolutions of C. L. Best's Model Sixty and Model Thirty tractors.

BEST Model A Sixty Track-Type Tractor to CATERPILLAR Model D8 Track-Type Tractor

The Best Model Sixty tractor S/N 101A debuted in June 1919 with engine characteristic references as follows: 6½" Bore, 8½" Stroke, 650 RPM. The machine was advertised and sold as a 60 H.P. machine. When Best's Model Sixty was tested in May 1921 at the Nebraska Test Lab, it received a rating of 56.33 H.P. These test results directed Best engineers in San Leandro to make engine changes that would establish a full 60 H.P. rating as advertised. Without any changes to the original bore, stroke, and engine RPM, and by only applying a series of minor adjustments to the cam shaft timing and rocker arm redesign, the Best Model Sixty tractor achieved a test rating of 77.10 H.P. in late 1924. As the horsepower of the Model Sixty tractor increased the supporting structure and drive line components of the machine were also enhanced. The last Model Sixty tractor built, S/N PA 13516, was completed at Caterpillar's Peoria, Illinois, facility November 25, 1931.

1. C. H. Wendell, *Nebraska Tractor Tests Since 1920* (Sarasota, FL: Crestline Publishing, 1985): 8.

From the first Diesel Model Sixty/Sixty-Five (serial number prefix 1C) tractor (with engine characteristic references D9900, 6⅛" bore, 9¼" stroke, 700 RPM) produced August 1, 1931, until the last machine was completed December 29, 1932, the proven reliable Model Sixty running gear was used. The gasoline-powered Model Sixty-Five (serial number prefix 2D) tractor (with engine characteristic references 9000G, 7" bore, 8½" stroke, 650 RPM) was manufactured from February 29, 1932, until February 19, 1933, also included the Model Sixty running gear. While the D9900 and 9000G series engines with their increased horsepower proved successful, the three-speed transmission drive line could not compete with the four- and five-speed transmissions offered by other manufacturers. To better compete in the track-type tractor market, Caterpillar engineers developed a heavier designed running gear with six-speed transmission to match up with the Diesel D9900 and Gas 9000G engines.

On February 6, 1933, Caterpillar Tractor Co. introduced the gasoline-powered Model Seventy Track-Type Tractor. This machine offered customers increased horsepower matching the Model 9500G gasoline engine (modified 9000G with 700 RPM) with the new six-speed transmission. The Model Seventy remained in production until February 18, 1937. The Diesel Model Seventy tractor was introduced with the D9900 engine with RPM increased to 820. This tractor had a short production schedule from serial number 3E1 February 8, 1933, to serial number 3E51 April 8, 1933.

On April 24, 1933, Caterpillar Tractor Co. introduced the Diesel-powered Model Seventy-Five Track-Type Tractor. The engine in this machine was a six-cylinder Diesel Model D11000 with characteristic references of 5¼" bore and 8" stroke and ran at 820 RPM. Because lubricating technology had not caught up with Diesel technology, it caused service issues for all modern high-speed Diesel engines. The well-designed Caterpillar running gear continued to live up to its expectations.

Introduction of the Model RD8 (serial number prefix 1H) series machine began with production starting in August 1935. The Diesel engine power unit for this machine was the Model D13000 with characteristic reference increased to 5¾" bore, same 8" stroke, and increased RPM to 850 as compared to previous Model D11000 used in the Model Seventy-Five tractor. Later called the Model D8 tractor, the engine and running gear for this configuration with minor modifications stayed in production until the mid-1950s.

Track-type tractors, which evolved from the Best Model Sixty to the Caterpillar Model D8, were built with one characteristic that remained constant throughout each model change: the bolt pattern used to mount attachments to the rear transmission case. Internal rear case structural designs were made over the years to compensate for engine speed and horsepower. However, a Best Sixty Belt Pulley or PTO from a tractor built in 1919 could physically be mounted to a D8 tractor built in the 1960s without modification to the bolt pattern.

BEST Model S Thirty Track-Type Tractor to CATERPILLAR Model D6 Track-Type Tractor

In February 1921 the first Best Model S Thirty Track-Type Tractor (serial number S999) left the San Leandro production facility. Number S999 was powered by a gasoline engine with characteristic references of 4¾" bore and 6½" stroke and operated at 800 RPM. This unit and number S1000 were used by the Best Company to demonstrate the versatility of this size tractor. The Model Thirty made its debut at the Nebraska Tractor Tests in May 1921 and did earn its rating of 30 H.P. Over the next three years, Best engineers made structural improvements and increased engine speed to 850 RPM. The Model Thirty, now renamed the Caterpillar Thirty, remained in production until October 4, 1932, by the Caterpillar Tractor Co.

In response to requests from customers for more machine power in the Model Thirty size tractor, the gasoline-powered Model Thirty-Five (serial number prefix 5C) series tractor went into production on February 29, 1932. The Model Thirty-Five featured characteristic referenced gasoline model 6000G engine with a 4⅞" bore and 6½" stroke at 850 RPM and was now offered with a four-speed transmission. Realizing that the Diesel engine was the power of the future, Caterpillar introduced the Diesel Model Thirty-Five (serial number prefix 6E) series on July 19, 1933. This tractor, powered by the model D6100 Diesel engine with characteristic reference 5¼" bore, 8" stroke operating at 850 RPM, was the first three-cylinder configuration Diesel engine offered by Caterpillar Tractor Co. The shorter physical length of this three-cylinder engine when compared with four- and six-cylinder engines necessitated a redesign of the Diesel starting engine. In October 1933 Nebraska Tractor Tests rated the Diesel Model Thirty-Five tractor at 46.15 H.P. Both the Model Thirty-Five 5C and 6E series tractors shared the same drive train and undercarriage design. On September 19, 1934, the final Model Thirty-Five gasoline tractor was produced. A little over one month later on October 23, 1934, the Model Thirty-Five Diesel tractor was produced on the production line in Peoria, Illinois.

The next track-type tractor in the evolution of the model appeared in late 1934. This was the Model Forty (serial number prefix 5G) powered by a gasoline model 6500G engine with characteristic reference 5⅛" bore, 8" stroke operating at 850 RPM. This change resulted in a substantial increase in horsepower to 51.53. During the same time period, the Diesel-powered Model Forty was offered as an option to the buying public. This tractor continued the use of the same engine (D6100) that powered the Diesel Model Thirty-Five but was now rated at 49 H.P. Both the gasoline and Diesel version of the Model Forty included an improved undercarriage and roller frame assembly enhancements over that of the previous Model Thirty-Five tractor.

One final design change was made to the line of Diesel engines that powered this model tractor. With the advent of reliable lubricating oils to support Diesel engine operation, Caterpillar engineers settled on a cylinder bore size of 5¾" while retaining the 8" stroke with full power at 850 RPM characteristic of the Model Thirty-Five tractor. This standardized Diesel engine configuration, designated model D6600, became the power source used in the new Model RD6 (later D6; serial number prefix 2H) series produced until 1941.

This historic model line of track-type tractors shared a similar design trait of the larger model tractors (the Best Model Sixty to the Caterpillar Model D8): the bolt pattern used to mount attachments to the rear of the transmission case assembly. Because engine speed remained at 850 RPM, by modifying the input shaft a PTO or Belt Pulley from a Model Thirty could be used on all model changes, including the R5 series, through the 2H series D6. Caterpillar engineers incorporated design features characteristic of the Best Model Thirty on tractor models ranging from the Model Ten to the Model D7 (serial number prefix 9G) tractors.

The definition of evolution is the gradual development of something into a more complex or better form. C. L. Best and his Best Tractor Co. engineering team in San Leandro set in motion the true meaning of evolution of the tracklaying tractor. This evolution continued with Caterpillar Tractor Co. and continues today with Caterpillar, Inc.

APPENDIX II

THE TALE OF TWO SEVENTY-FIVES

Tractors are sometimes known by the horsepower they are capable of producing. This horsepower translates into what type of work the tractor was designed to perform. The lower the horsepower, the smaller the machine and the lighter the job load it can perform. Likewise, a greater horsepower results in a larger tractor capable of heavier duties. Through the years, tractor manufacturers have sized their machines to compete for the same market share resulting in the same horsepower designation. But just because those numbers are the same, it doesn't follow that the tractors are copies of each other. The following table compares the Best 75 H.P. Tracklayer to the Holt 75 H.P. Caterpillar.

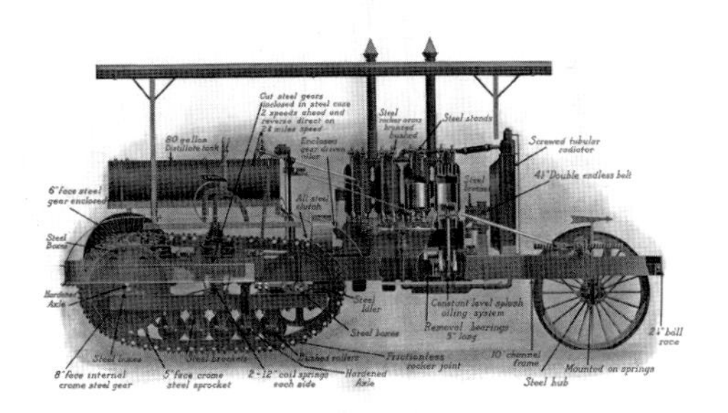

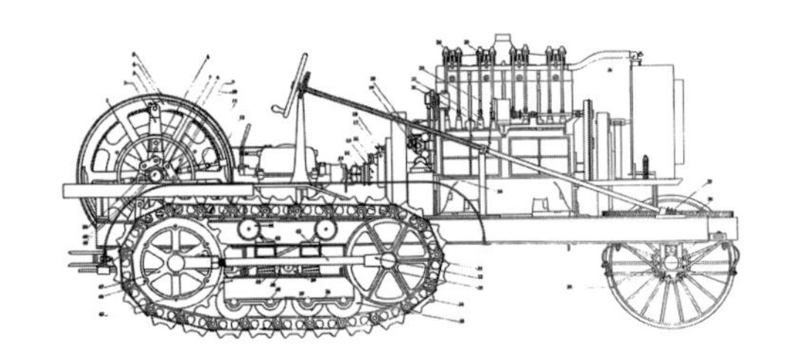

	Best 75 H.P. Tracklayer	**Holt 75 H.P. Caterpillar**
Engine:	7¾×9 4 cylinder 1,698.2 cubic inches 75 H.P. at 450 RPM Flywheel secured by bolts to forged crankshaft flange	7½×8 4 cylinder 1,413.7 cubic inches 75 H.P. at 550 RPM Flywheel secured by bolts and keyed to crankshaft
Flywheel Clutch:	Best-designed flywheel clutch	Holt-designed flywheel clutch
Transmission:	Best-designed transmission Two forward speeds: 2.375 MPH and 1.50 MPH One reverse speed: 1.625 MPH	Holt-designed transmission Two forward speeds: 2.93 MPH and 2.18 MPH One reverse speed: 2.18 MPH
Steering:	Front tiller wheel Power steering Differential Independent brake for each track	Front tiller wheel Clutch and brake system
Transmission of Power to Rear Sprockets:	Shaft from differential ending with a pinion gear that meshed with the internal gear of the track sprocket	Steering clutch shaft with a sprocket on each end that runs a roller chain to the track sprocket shaft
Roller Frame:	Oscillating	Fixed
Tractor Weight:	28,000 pounds	22,860 pounds
Tractor Length (L), Width (W), Height (H):	L 22'4"; W 8'7"; H 10'	L 19'; W 8'8"; H 10'
Track Gauge:	70"	80"

APPENDIX III

TIME LINE

1838 › Daniel Isaacs Best is born to John George and Rebecca Best in Ohio.

1847 › John George Best moves his family to the Vincennes, Iowa, area.

1852 › Samuel Best departs Iowa for California.

1859 › Daniel Best travels from Iowa to Fort Walla Walla, Washington.

1862 › Henry and Lavina Best leave Iowa and settle near Yuba City, California.

1869 › Daniel Best visits his brother Henry in California and stays.

1870 › Daniel Best builds three successful grain separators.

1872 › Daniel Best marries Meta Steinkamp.

1875 › Daniel and Meta Best move to Auburn, Oregon.

1878 › Clarence Leo Best is born to Daniel and Meta Best.

1885 › Daniel Best founds the Daniel Best Agricultural Works in San Leandro, California.

1886 › Daniel Best moves his family from Oregon to San Leandro.

1887 › Daniel Best applies for a patent for his combined harvester.

1888 › Oregon-based Marquis deLafayette Remington brings his steam-traction engine to San Leandro.

› Daniel Best buys the rights to manufacture the Remington engine on most of the West Coast.

› Daniel Best first attempts to build a steam-powered combined harvester.

1889 › First Best steam-traction engine sells for $4,500.

1891 › Daniel Best applies for a patent for a single-cylinder horizontal gas engine.

1893 › Best Manufacturing Co. (BMC) is incorporated.

1894 › Meta Best dies.

1896 › Daniel Best applies for a patent for an oil-vapor traction engine.

1897 › C. L. Best is made buyer for BMC.

1898 › C. L. Best is superintendent for BMC.

› BMC produces a prototype horseless carriage.

1900 › C. L. Best marries Pearl Margaret Gray.

1901 › Alvin O. Lombard of Waterville, Maine, is granted a patent for a track-type engine and soon begins commercial production.

1903 › Harvey Beckwith, from Oakland, California, applies for a patent for a self-laying track-type device.

1904 › BMC files a patent infringement case against the Holt Manufacturing Company (HMC).

› Benjamin Holt demonstrates his paddle wheel traction engine.

1907 › BMC receives a favorable judgment of $35,000 against HMC.

› C. L. Best begins investigating the self-laying track principle.

› Elizabeth (Betty) Best is born to C. L. and Pearl Best.

1908 › HMC appeals the judgment against it.

› October 8: HMC purchases BMC.

› C. L. Best elected president of BMC.

1909 › Holt-owned BMC shows a $90,000 profit for fiscal year 1909.

1910 › January 15: C. L. Best resigns as president and superintendent of BMC.

- › March 5: C.L. Best Gas Traction Co. (CLBGT) is incorporated.
- › CLBGT begins manufacturing round-wheel tractors at its Elmhurst plant.

1911 › C.L. Best Steel Casting Co. is in production.
- › CLBGT builds a ball-race track-laying tractor.
- › August: C. L. Best shifts the track design to bearing rollers.

1912 › November: Daniel Gray Best is born to C. L. and Pearl Best.
- › December: CLBGT's new 70 H.P. Track engine, featuring rocker joint link pins, is advertised.

1913 › BMC in San Leandro is permanently shut down.
- › Oscar Starr becomes a CLBGT employee.

1914 › 75 H.P. Tracklayer is introduced.

1915 › Philip Rose issues the Rose Report on the products and corporate health of CLBGT and HMC.
- › February 19: HMC files a patent infringement suit against CLBGT.
- › February 20 through December 4: Panama Pacific International Exposition is held in San Francisco.
- › On legal advice, C. L. Best purchases the track-laying patents from A. O. Lombard and sues HMC for infringement of those patents.

1916 › CLBGT moves production from Elmhurst to San Leandro on the site of the former Best Manufacturing plant.

1917 › March: C. A. Hawkins buys 51 percent of CLBGT stock and becomes company president with C. L. Best as vice president.
- › Oscar Starr leaves CLBGT for the Holt plant in Peoria, Illinois.

1918 › March: C. L. Best takes back control of CLBGT.
- › December: Hawkins's share of CLBGT stock is bought out, and Hawkins retires from the business.
- › December 30: A compromise settlement is reached between CLBGT and HMC. The court declares the Holt patent is valid and CLBGT did infringe upon it. Best sells his patents to Holt for $200,000 and has to pay the HMC a royalty for each tractor it produces. HMC now has to defend its patents against other manufacturers.

1919 › CLBGT signature tractor Model A Sixty goes into production.
- › C. L. Best assembles a management team that will remain together until 1951: Harry H. Fair, B. C. Heacock, R. C. Force, and Oscar Starr.

1920 › U.S. economy struggles.
- › October 15: CLBGT changes its name to the C.L. Best Tractor Co. (CLBT).
- › December 5: Benjamin Holt, founder of HMC, dies.

1921 › CLBT Model S Thirty tractor goes into production.
- › Severe economic conditions continue to plague CLBT and other manufacturers.

1923 › August 22: Daniel Best dies after a short illness.
- › CLBT has a profitable year.

1924 › The number of CLBT tractor dealerships increase.
- › CLBT shows a net earnings increase of nearly 30 percent over 1923.

1925 › Financial markets are alluding to important developments concerning CLBT and HMC.
- › February 28: *Stockton Independent* headline reports that "Holt Company Control Bought by C.L. Best Tractor heads."
- › March 10: *Stockton Independent* reports that C. L. Best and other Best Company officials inspect the Holt factory in Stockton, California.
- › April 15: The Caterpillar Tractor Co. is incorporated under California laws.
- › April 17: The executive management team for the Caterpillar Tractor Co. is announced with C. L. Best as chairman of the board. Most board members are from the Best management team, with only two members representing HMC.
- › May 14: Per stockholders' votes, the C.L. Best Tractor Co. and HMC cease existence and all plants,

businesses, assets, and liabilities are assumed by the Caterpillar Tractor Co. San Leandro, California, becomes the corporate office for the new company.

› December: The Western Harvester Company, with all its stock owned by Caterpillar, opens in Stockton.

1926 › C. L. Best begins the development of practical Diesel power for tractors.

1928 › The Russell Grader Mfg. Co. of Minneapolis, Minnesota, is purchased by the Caterpillar Tractor Co.

1929 › October 29: The Stock Market crashes on Black Tuesday and signals the beginning of the Great Depression.

1930 › June: The first Caterpillar-designed and -built Diesel engine is completed.

› July 30: Caterpillar announces the end of tractor production in San Leandro with tractor manufacturing now exclusively at the Peoria plant. San Leandro will host the research division of the company.

› August: A divorce decree is granted to Pearl Margaret Best, ending her marriage to C. L. Best.

1931 › August 29: C. L. Best marries Irene Wagness Hansen.

› August 31: The Caterpillar Diesel Sixty Tractor is announced.

1932 › The Caterpillar Tractor Co. posts a loss of $1.6 million.

1933 › September 23: Caterpillar Tractor Co. produces its one thousandth Diesel engine.

1934 › November 5: The five thousandth Caterpillar Diesel engine is produced.

1935 › November 13: The Peoria plant celebrates the ten thousandth Diesel engine produced. The Caterpillar Company has now produced 640,450 total Diesel horsepower.

1951 › September 22: C. L. Best dies at the age of seventy-three. The executive committee that was formed in 1919 at the CLBGT and continued with the Caterpillar Tractor Co. loses its first member.

APPENDIX IV

ARMY IMAGES

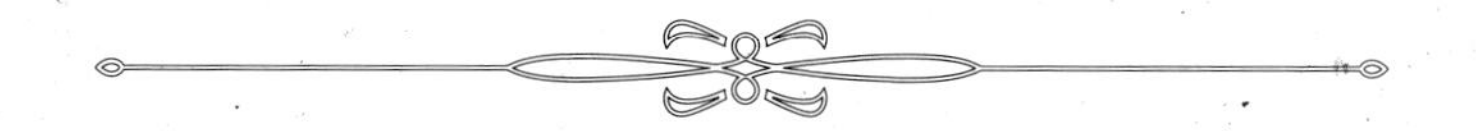

(Top left) *A Caterpillar Ten hauling a twelve-passenger, three-motored Fokker plane from its hanger at the Oakland, California, airport. This plane was used in service between Oakland and Los Angeles, California. September 1928.* (Top right) *Military officers pose for a photo with Caterpillar Thirty and Sixty tractors.* (Bottom right) *Caterpillar Sixty tractor towing a five-inch Seacoast gun on a mobile carriage. April 18, 1928.* (Bottom left) *Caterpillar Thirty tractor emerging from a mud hole. June 23, 1927.* Photos courtesy of Col. O.P. Winningstad, U.S. Army.

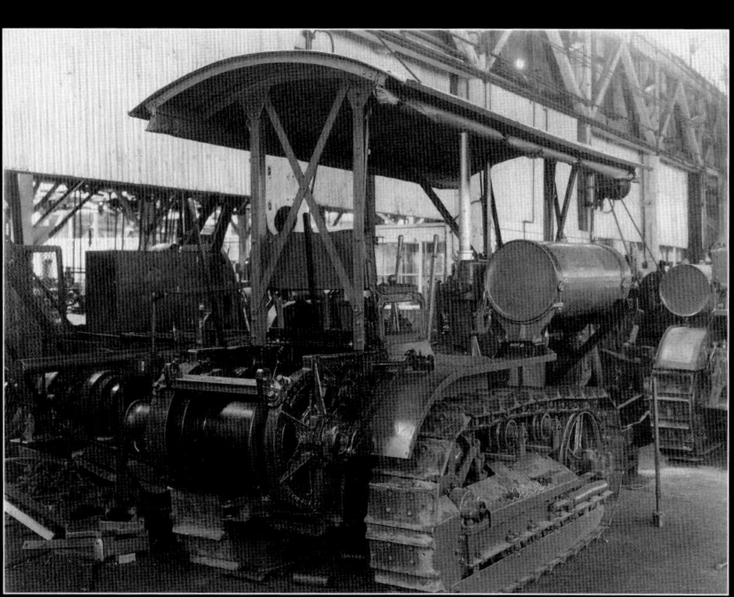

(Top) *Best Sixty tractor climbing over a two-foot log. March 2, 1925.* (Bottom right) *The Swedish Army uses a Caterpillar Thirty tractor to pull heavy artillery. Tests were conducted in the outskirts of Stockholm. November 8, 1929.* (Bottom left) *Willamette winch on a Caterpillar Thirty tractor. January 29, 1927.* Photos courtesy of Col. O.P. Winningstad, U.S. Army.

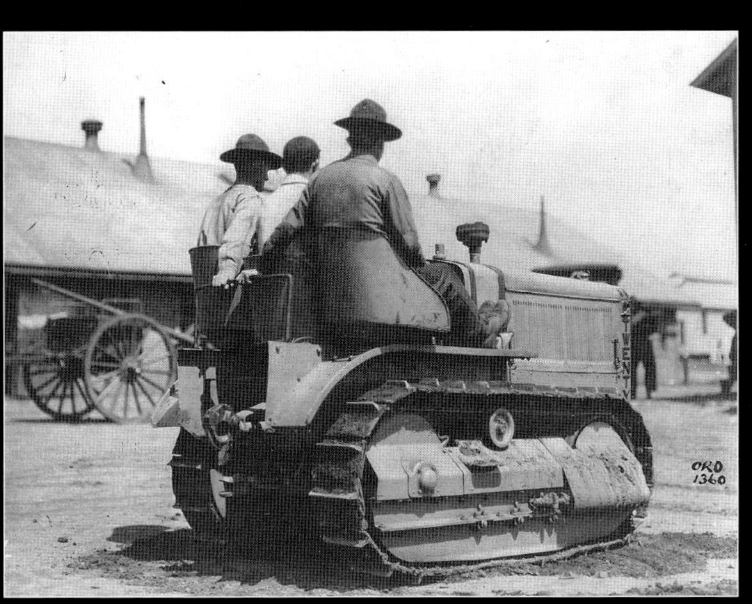

(Top) *A Best Sixty cruiser-type is shown during the fording tests at Woodpecker Point Beach. January 14, 1925.* (Bottom right) *Side view of a Caterpillar Twenty tractor with three seats.* (Bottom left) *Rear view of a Caterpillar Twenty tractor with three seats.* Photos courtesy of Col. O.P. Winningstad, U.S. Army.

APPENDIX V

C.L. BEST *CATALOGUE "F"*

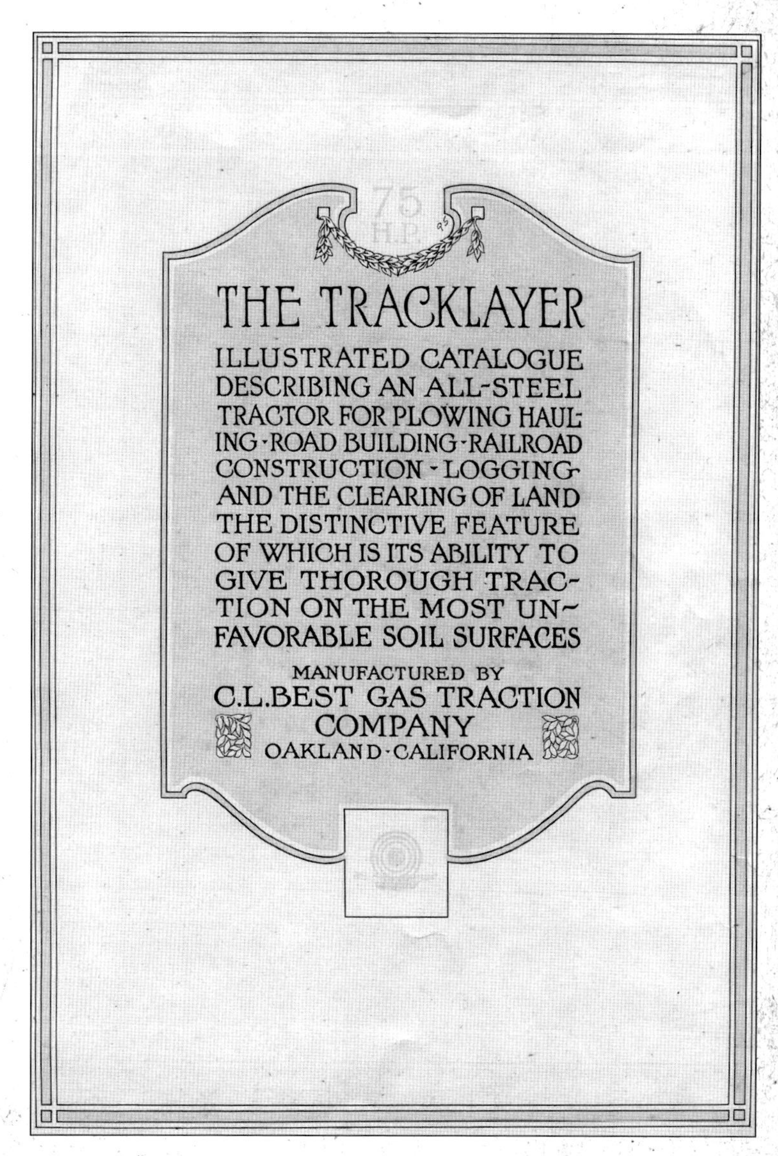

THE TRACKLAYER

ILLUSTRATED CATALOGUE
DESCRIBING AN ALL-STEEL
TRACTOR FOR PLOWING HAUL-
ING · ROAD BUILDING · RAILROAD
CONSTRUCTION · LOGGING
AND THE CLEARING OF LAND
THE DISTINCTIVE FEATURE
OF WHICH IS ITS ABILITY TO
GIVE THOROUGH TRAC-
TION ON THE MOST UN-
FAVORABLE SOIL SURFACES

MANUFACTURED BY
C.L.BEST GAS TRACTION
COMPANY
OAKLAND · CALIFORNIA

The reeds or tules in this picture show the marshy nature of the soil. Neither horses nor round-wheel tractors can plow such land. Outfit of E. E. Brownell, Suisun, California.

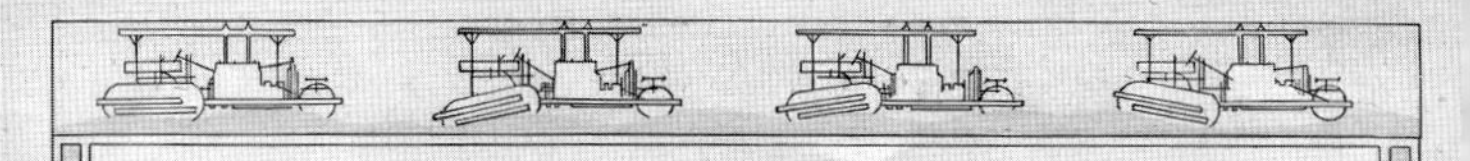

HORSE SENSE

The progress of the tractor industry has never ceased. Steadily, each year, tractors have increased by large numbers. Factories have grown in size until many times larger than in their beginnings. All over North America and in notable numbers in South America and Continental Europe the farmers have gradually accepted the iron horse as their greatest gift from modern mechanicians.

To say that its use was eagerly accepted would be untrue. Every man who has learned to love horses cannot witness their passing without a twinge of regret. He does not give them up for a strange creature of steel without much hesitancy and mental forebodings. No such wonderful change as the automobile has wrought in the cities has yet occurred on the farms.

But the cry of the age is for efficiency. The world demands it. The United States could not feed itself and a large part of the world if our farmers still used hand-reapers and the walking plow. One hundred years ago the families of four farms supported one in the cities. Now they support three in the cities with a substantial surplus for export. Modern machinery, making the cultivation of larger areas and the better cultivation of smaller areas possible has brought the change.

Other changes have come. The farmer is no longer a drudge. He is a capitalist, with widespread interests. He asks for freedom from unnecessary labor. Each horse requires twenty-seven minutes of someone's care each day. There are thirty million work-horses and mules in the United States.

What a terrific waste of human energy! The time spent each day in caring for one horse, if spent on a tractor equal to thirty horses, will keep it in the best of condition all the year round.

A horse can work but six hours a day steadily. A tractor can work twenty-four and lives longer than the horse in hours of work. There is no bulk to its fuel. The machine and a year's supply will occupy but a tenth of the space necessary for animals of the same power and their feed. And the yield of five acres, which each horse eats annually, is withheld from human consumption to feed to them.

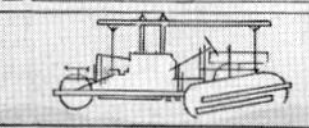

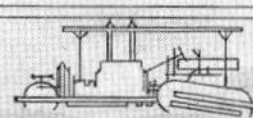

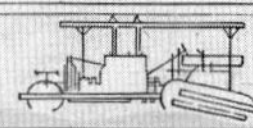

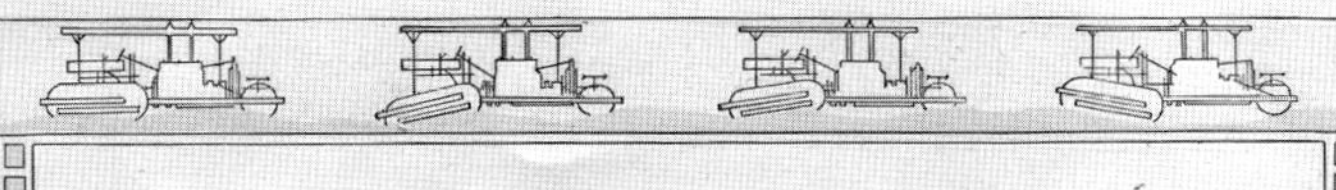

C. L. Best 75 horse-power Tracklayer, fully equipped, electrically lighted, self-steering.

Labor is scarcer and wages higher. Two men on a tractor and plows can do as much work in a day as five men with thirty-five horses. And the tractor will often combine two or more operations.

Horses can no longer compete economically. They must give way to machinery, as both men and horses have done in all other classes of human endeavor.

But still another reason than man's inherent love for his friend the horse has delayed the general adoption of the tractor. The first tractors were built with large, round drive wheels incapable of good traction in unfavorable conditions.

The wheels slipped, mired and could not pull in rainy weather and in sandy or marshy soil. They failed in many ways to assume the work formerly done by the animals.

An improvement was imperative. Many attempts were made over long periods of time. "Walking wheels," "pedrails," crude "platform wheels" all had their day. Finally out of the chaos of failure came the big success.

The remarkable and immediate popularity of the Best Tracklayer has astounded many people, but to those who are experienced in tractor operation the causes are at once apparent. The round-wheel tractor is limited in its operation. It cannot work unless soil conditions are favorable, unless

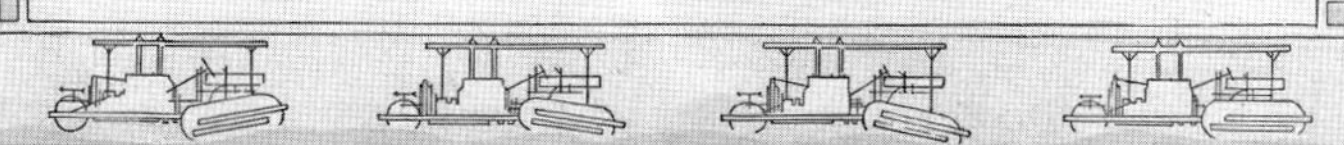

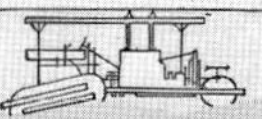 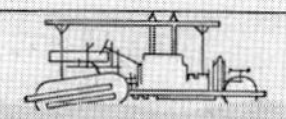 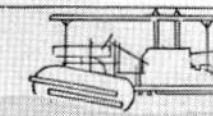

the huge drive-wheels can secure sufficient grip on the surface of the ground to propel the outfit.

Since wholly favorable operating conditions are the exception rather than the rule and the majority of farms composed of several kinds of soil, both good and bad, any improvement which minimizes the loss of time and makes tractor operation as dependable as that of horses is certain to meet with much encouragement.

The theory of bringing a larger propelling surface in contact with the ground surface and thus increasing both the grip and the sustaining power is the only correct one. As an illustration of this we have the board which supports a man's weight on the softest mudhole.

This in effect is precisely what the Best Tracklayer tracks accomplish. The tracks possess sufficient ground contact area to support the weight of the tractor on ground so soft that horses cannot walk thereon. Such an area also gives a gripping surface which prevents all slipping.

In all farm work, therefore, and in road work, construction work, logging and clearing land, a tractor has been provided which will work in any weather or soil at all seasons of the year. The Best Tracklayer knows no idle moments when work is needed.

Martin Lund of Stockton, California, is hauling forty tons of grain in this picture. Best Tracklayer road work is one of its most useful purposes. Winter or summer, the condition of the roads need not be considered.

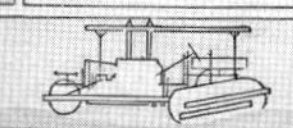 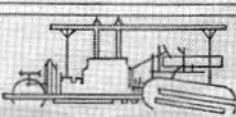 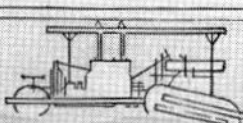 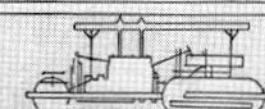

BEST TRACKLAYER INFLUENCE

In every industry, since the beginning of time, someone always improves on the work of predecessors. No single individual has yet succeeded in producing the last word in anything. Watt invented the steam engine: Stephenson put it to useful work in the first locomotive, and Fulton the first steamboat. Franklin discovered electricity: Edison gave it to the world.

There are those who claim the platform wheel to be their own exclusive creation and attack all others who have since developed it. As a matter of fact, there are no basic patents on the platform wheel. The first record of its practical use comes to us from the lumber camps of New England, where gigantic steam tractors were so equipped for hauling logs over the snow in winter time. That it was known and used for various purposes in England and other European countries even before that time, is definitely established. Captain Scott, in his last great dash for the South Pole, carried sleds equipped with revolving tracks. The historic diary, recovered by the

This, and the illustration on the opposite page, picture the remarkable performance of a Best Tracklayer owned by H. J. Knowles, Winters, California. Loaded, the tractor forded a stream without difficulty.

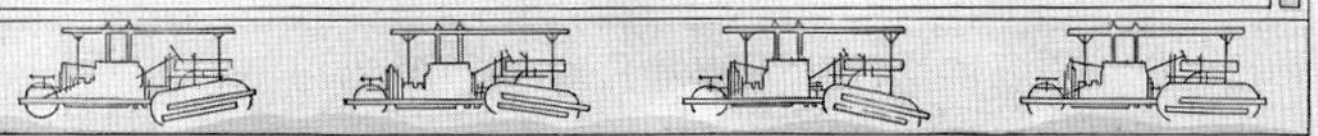

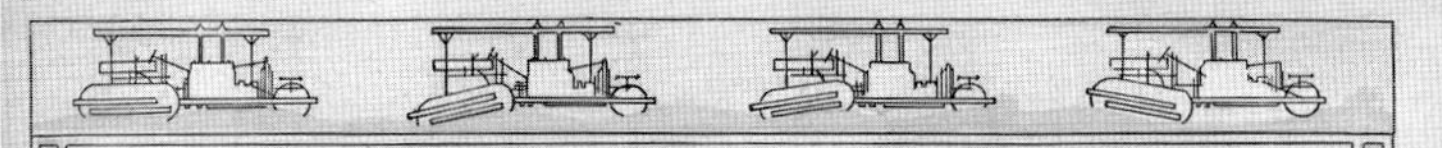

If the Tracklayer can, when necessity requires, do such work as this, can there be any doubt of its advantages over round-wheel tractors?

searching party after his death, refers to these sleds as complete successes in tractive effect although ultimate failures because the excessive cold cracked the cylinders of the motors.

The Best Tracklayer, because of its position as the largest competitor in the field, receives the most attention from those who shriek "infringement," "imitation," and seek to cause timid buyers to look elsewhere. But no infringement suit has ever been attempted and never will be. Nor can a thing which is superior be called an imitation. Would one call the automobile of today an imitation of those of ten years ago?

As Mr. Best has said in his letter on later pages, the Best Tracklayer was built and the price reckoned when the machine was complete. A cheap tractor should never be built or bought. It has no justification in any kind of work. A tractor is a heavy-duty machine. Its work is hard bolt-straining, steel-twisting pulls. When attached to a load within its horsepower every piece of iron, steel, brass or copper should be able to bear its share of the work without danger of breakage.

Who ever heard of railroads buying cheap locomotives? They buy them either large or small but never cheap. A tractor is a locomotive without a roadbed. It hauls plows and harvesters instead of coaches. It must be built to resist terrific strains.

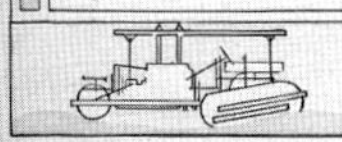
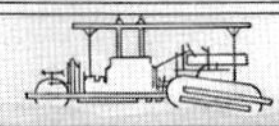
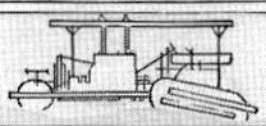
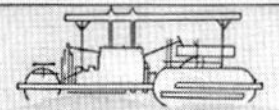

Every day counts when in the harvest field. The tractor which does its work without delays from breakage is the one you should buy. Schleuter & Company, Orland, California, own this outfit.

What shall we say, then, of the tractor builder who continues, year after year, to send out "cast iron" tractors? He knows full well that only the best of steel and the strongest and most careful construction can give a return on the money invested. Can there be any excuse for such a policy?

We did not think so when we began the construction of the Tracklayer. The farmers of California had been burdened with "semi-steel" tractors for years. Extra bills were appalling. Breakdowns were continuous. Tracks wore out in a few weeks. Complaints were loud and long but without avail. No relief whatever was given until the Best Tracklayer appeared.

For a complete understanding of the exact situation we refer you to the files of your California farm journals for the latter part of 1913. In these issues appeared the first announcements of the Best Tracklayer. Along with them you will find other tractor advertising. Machines which had changed but little since their introduction were suddenly discovered to require a most amazing number of improvements. Old frames were entirely discarded. New tracks took the place of old. Axles, sprockets, transmission, motors—all were changed almost over night.

The Best Tracklayer, built of steel and built to do its work as it should be done, had forced the issue.

But it requires real proof to establish such claims as these.

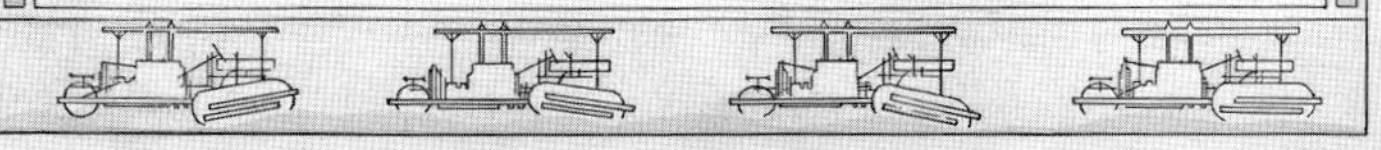

To simply assert that this or that is true, padding the statement with meaningless generalities, high-sounding phrases and superlative descriptions no longer serves to convince those who think clearly.

We are fortunate in having such proof. In fact, such figures as we shall now show have never to our knowledge been displayed by any tractor manufacturer; which makes their value all the greater.

On page 32, in the carefully kept records of the California Fruit Canners Association, who own the Marin Meadows Farm and two Best Tracklayers, you will find the interesting and unusually convincing fact that this institution figures the *percentage of efficiency* demonstrated by their tractors.

Most people keep no records whatever. The best of them show only receipts and disbursements. But here we have the *percentage of efficiency* and it shows *that the Best Tracklayer, out of 747 operating hours, only 17.5 hours' delay was caused by tractor trouble, or precisely 98 per cent tractor efficiency.*

PROMINENT TRACKLAYER OWNERS

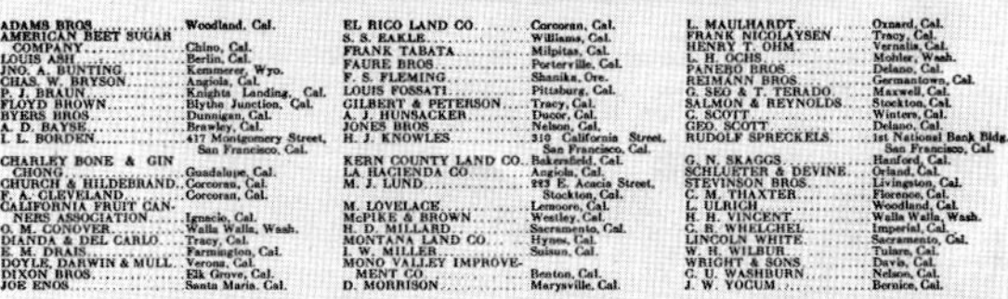

ADAMS BROS	Woodland, Cal.
AMERICAN BEET SUGAR COMPANY	Chino, Cal.
LOUIS ASH	Berlin, Cal.
JNO. A. BUNTING	Kemmerer, Wyo.
CHAS. W. BRYSON	Angiola, Cal.
P. J. BRAUN	Knights Landing, Cal.
FLOYD BROWN	Blythe Junction, Cal.
BYERS BROS	Dunnigan, Cal.
A. D. BAYSE	Brawley, Cal.
I. L. BORDEN	417 Montgomery Street, San Francisco, Cal.
CHARLEY BONE & GIN CHONG	Guadalupe, Cal.
CHURCH & HILDEBRAND	Corcoran, Cal.
F. A. CLEVELAND	Corcoran, Cal.
CALIFORNIA FRUIT CANNERS ASSOCIATION	Ignacio, Cal.
O. M. CONOVER	Walla Walla, Wash.
DIANDA & DEL CARLO	Tracy, Cal.
E. M. DRAIS	Farmington, Cal.
DOYLE, DARWIN & MULL	Verona, Cal.
DIXON BROS	Elk Grove, Cal.
JOE ENOS	Santa Maria, Cal.
EL RICO LAND CO	Corcoran, Cal.
S. S. EAKLE	Williams, Cal.
FRANK TABATA	Milpitas, Cal.
FAURE BROS	Porterville, Cal.
F. S. FLEMING	Shaniko, Ore.
LOUIS FOSSATI	Pittsburg, Cal.
GILBERT & PETERSON	Tracy, Cal.
A. J. HUNSACKER	Ducor, Cal.
JONES BROS	Nelson, Cal.
H. J. KNOWLES	310 California Street, San Francisco, Cal.
KERN COUNTY LAND CO	Bakersfield, Cal.
LA HACIENDA CO	Angiola, Cal.
M. J. LUND	223 E. Acacia Street, Stockton, Cal.
M. LOVELACE	Lemoore, Cal.
McPIKE & BROWN	Westley, Cal.
H. D. MILLARD	Sacramento, Cal.
MONTANA LAND CO	Hynes, Cal.
I. W. MILLER	Suisun, Cal.
MONO VALLEY IMPROVEMENT CO	Benton, Cal.
D. MORRISON	Marysville, Cal.
L. MAULHARDT	Oxnard, Cal.
FRANK NICOLAYSEN	Tracy, Cal.
HENRY T. OHM	Vernalis, Cal.
L. H. OCHS	Mohler, Wash.
PANERO BROS	Delano, Cal.
REIMANN BROS	Germantown, Cal.
G. SEO & T. TERADO	Maxwell, Cal.
SALMON & REYNOLDS	Stockton, Cal.
C. SCOTT	Winters, Cal.
GEO. SCOTT	Delano, Cal.
RUDOLF SPRECKELS	1st National Bank Bldg., San Francisco, Cal.
G. N. SKAGGS	Hanford, Cal.
SCHLUETER & DEVINE	Orland, Cal.
STEVINSON BROS	Livingston, Cal.
C. M. THAXTER	Florence, Cal.
L. ULRICH	Woodland, Cal.
H. H. VINCENT	Walla Walla, Wash.
C. R. WHELCHEL	Imperial, Cal.
LINCOLN WHITE	Sacramento, Cal.
W. H. WILBUR	Tulare, Cal.
WRIGHT & SONS	Davis, Cal.
C. U. WASHBURN	Nelson, Cal.
J. W. YOCUM	Bernice, Cal.

One hundred and six inches of plows eight inches deep in black adobe.

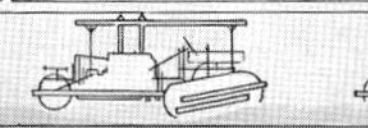

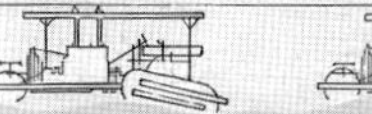

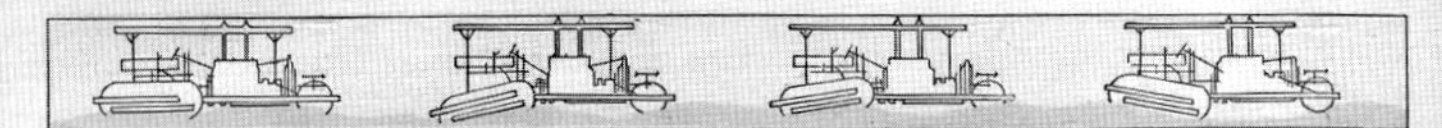

THE MEANING OF MY NEW GUARANTEE
by C. L. BEST

Over my own signature I wish to announce a new guarantee for C. L. Best Tracklayer tractors. There are so many unfounded and extravagant assertions made for tractors that such a course is necessary for those who are really doing what they claim. But I also want it to be a straightforward declaration establishing the status of the C. L. Best Tracklayer among its competitors. I want it to indicate that I am building a tractor into the construction of which the question of price does not enter.

There are four or five names in the automobile industry representing the highest type of construction. Even as you read this those names have entered your mind. Their worth is firmly established. Price is never considered when they are planned and built. In this way have I built my Tracklayer and I intend to so make it known. I not only claim such distinction; I am going to back it up with a guarantee that would ruin me if the Tracklayer were not built of the best material available and workmanship of the highest type.

Do you have hills? If so, you can only depend on the Best Tracklayer, of all tractors, to do your work. This machine is owned by the Union Sugar Company, Betteravia, California.

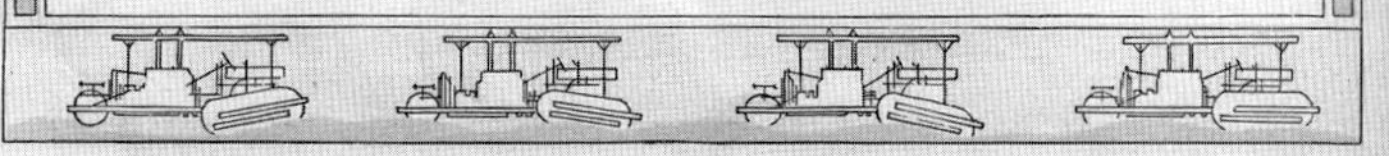

Another piece of marshy plowing showing ten fourteen-inch bottoms.

I determined, when I first started to build a track engine, to give the farmers of the West a tractor worth every cent they paid for it. Up to that time it was generally supposed that a track engine could not be built to run economically. I thought otherwise. I knew I could build a similar tractor and build it right. I know that proper construction and good materials will produce a good machine providing the basic principle is right. I was satisfied the basic principle involved in the revolving track was better than the round wheel and that good material and right construction was all that was needed to make this type of machine a success.

I commenced by using steel throughout the tractor—and good steel. No mushy cast iron was used. Where iron was practicable I used close-grained hard iron. Where iron was too weak I did not hesitate to use steel. All the gears, sprockets, rollers and track chains are composed of forty-point carbon steel; all boxes, brackets, guides, rocker arms and stands, all parts of the clutch, the fan, fan spider and fan bracket are made of twenty-five point steel. Even the steering wheel and hubs are made of twenty-five point steel. No tractor in the world contains so much steel so well and judiciously used. Manganese, chrome, chrome nickel, vanadium—all were used where best fitted to perform some particular work.

And I used steel-geared transmission—not chains. Any mechanical engineer knows the gear drive to be far better than chains. There isn't

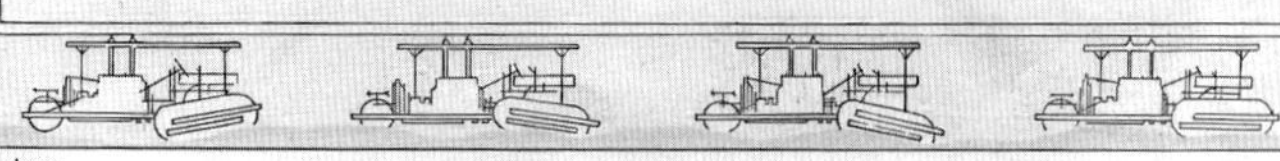

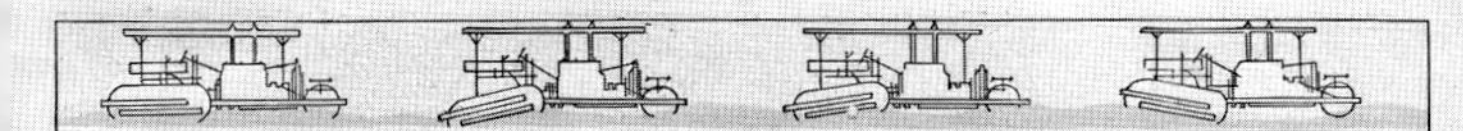

even an argument about it. And if any one tells you they can't find steel hard enough to use in heavy-duty tractor gears and must therefore use chains (the usual excuse), don't you believe it! It's totally untrue. The real answer is that the gear drive as I have it in my Tracklayer adds about three hundred dollars to the manufacturing cost. It's expensive but it lasts. At that, I make these gears myself in my own Bessemer steel foundry. I am the only tractor manufacturer on the Pacific Coast owning and operating a Bessemer steel plant. I make all my own steel parts. I don't buy any of them from some one else.

This policy and the Best Tracklayer both proved to be right from the beginning. The first machine was better than others that had been on the market for years. You probably remember the noticeable scramble to patch up old construction as soon as my machine appeared.

Thus the Best Tracklayer immediately became a pacemaker and has remained so. It met with quick appreciation. Our business this year was three times greater than last year. It is going to be much greater next year.

But to carry conviction with my claims I am going to issue this new guarantee. I say that my Tracklayers are durable; that they are built to last. I say that materials and construction have never been slighted and cheap metals and methods never used. I say that I have striven from the

Soft, ashy soils cause most of the tractor trouble in the harvest field. The Best Tracklayer, with its enormous gripping surface can always be depended upon. Mack Lovelace, Lemoore, California, owner of this outfit, will confirm this.

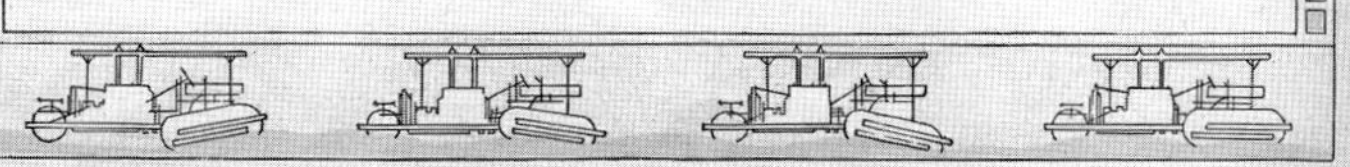

Rudolph Spreckels of San Francisco owns a Tracklayer. In such work as this he finds it unequaled.

beginning to make the Best Tracklayer known as highest possible type of mechanical farm power. I say that I have succeeded in doing so.

But more than that, I want you to believe in me personally; I want you to think I can be relied upon; I want you to have confidence in me as a manufacturer. I know how to build tractors. My father before me built them. He taught me how and educated me for that purpose. From boyhood I have been in no other business. I am specializing on them now.

So I am issuing the following guarantee. It backs up all I have said:

> "The C. L. Best Gas Traction Company guarantees this Tracklayer tractor to develop seventy-five (75) brake horsepower at 450 R. P. M. and that it will under ordinary operating conditions do the work of thirty-five (35) average horses.
>
> [NOTE.—In the case of the thirty horsepower Tracklayer the brake horsepower is guaranteed at thirty and the drawbar horsepower at sixteen. Other than this, there is no change in the text.]
>
> "The C. L. Best Gas Traction Company hereby guarantee the Tracklayer gas tractor sold under this contract to be built in a workmanlike manner of substantial material free from defects, and if within one year from date of delivery any defects appear in any of the parts, such parts will be furnished free F. O. B. Oakland, California, provided such parts have not been injured by misuse or neglect.

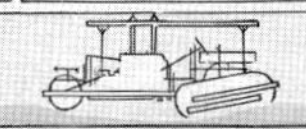
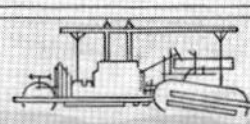

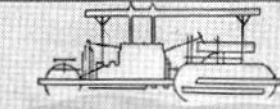

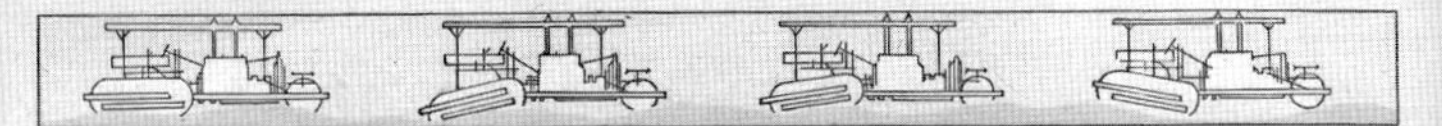

"That the purchaser may be protected against excessive wear of the tracks, the C. L. Best Gas Traction Company hereby agrees to accept the payment of fifty dollars ($50) at the time of purchase in addition to the regular list price of the tractor, for which the C. L. Best Gas Traction Company hereby agrees to repair said tracks free of charge at any time during one year after date of delivery should said tracks need repair either once or more than once during that period."

The first part of the guarantee covers the brake and drawbar horsepower. The second part covers the construction of the entire engine and guarantees you against defective material and construction. The third part substantially guarantees you against excessive upkeep of the track.

We don't nor never have claimed the tracks to be free from upkeep expense. They do cost something in wear and tear—about fifty dollars a year. They are worth ten times that to any farmer because of the additional work you get out of the Best Tracklayer compared to a round-wheel tractor. But I do claim that my rocker-jointed steel track should not cost more than fifty dollars a year in upkeep. Some cost one hundred and fifty dollars a year; mine don't, and I am willing to back up my claims.

I am, therefore, agreeing to accept fifty dollars extra at the time you buy the Best Tracklayer and in return if the tracks need repairs in less than

Another convincing demonstration of Tracklayer power. This tractor is hauling forty tons of rock on highway construction in California.

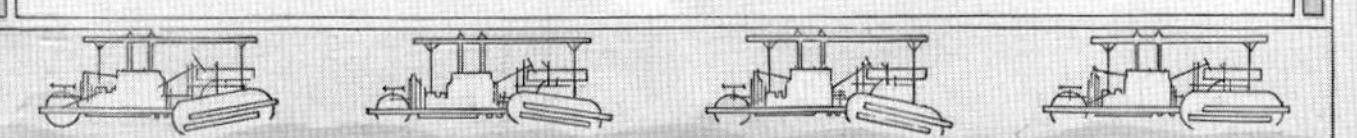

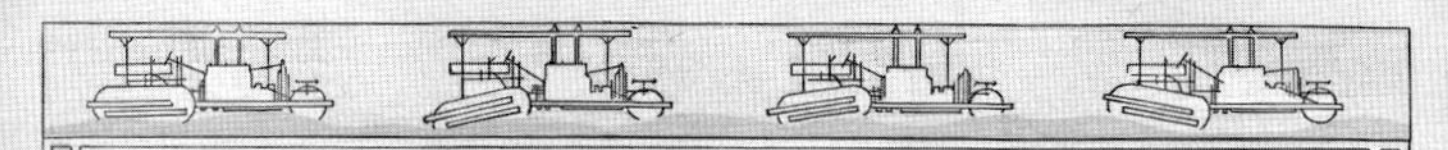

a year—as many times as they need repairs—I'll repair them free of charge.

I believe this guarantee gives you all the protection you need. If you can point out a weakness in it, do so. I want you to be satisfied.

The new thirty horsepower Best Tracklayer I am announcing on another page is a good little machine. I have built it carefully of the same substantial material which makes the seventy stand out above its competitors. A long time has been spent searching for its weak points and I have put it through the hardest kind of tests. It is right in every particular.

There may be some things in this statement you don't understand. I may not have made myself clear. If so, I shall be only too glad to see you personally and explain anything you want to know or answer any of your letters. I am always glad to reply to any number of questions.

I should like also to have you come and see my factory and the workmanlike way we build the Best Tracklayer. An investment of the amount of money involved when you buy a tractor demands the closest investigation on your part. Come and see us before you make up your mind which machine to buy.

Sincerely yours,

C. L. Best

President C. L. Best Gas Traction Company.

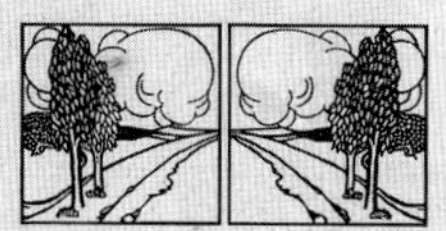

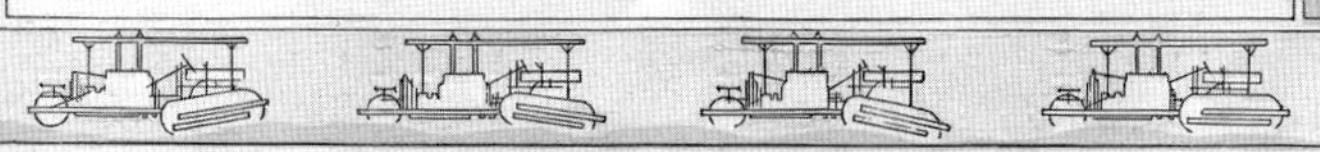

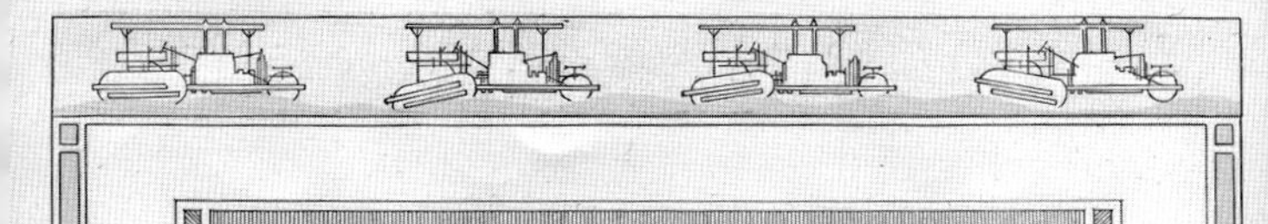

THE MOTOR

Not many difficulties are encountered today in the construction of a heavy-duty tractor motor. Their excellence is largely a matter of the correct blending of certain features found in other types, that the two important points, accessibility and dust protection, may be realized. These we have achieved to a most satisfactory degree in the Best Tracklayer motor. Then comes care in construction and selection of materials, and it is in these latter points that the difference between one and another becomes apparent. So long as a motor will run and produce power many manufacturers are content. Some have even gone so far as to place straight-head cylinders on "L" head bases, a large number of the latter having been left on hand because of the failure of the "L" head type. The result,

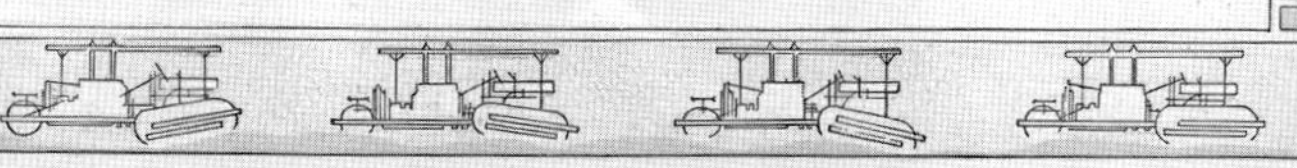

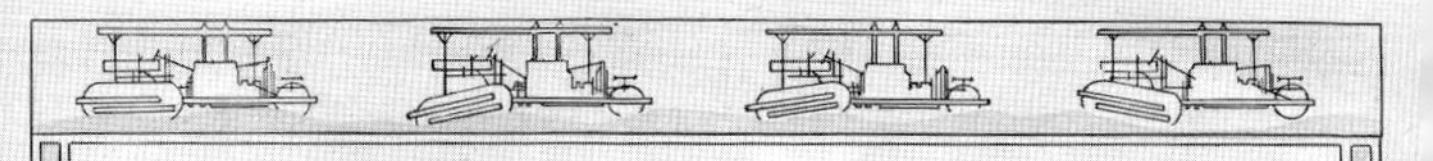

The mechanical oiler is gear driven and accessible.

a queer "dish-face" motor, the valve rods of which slanted inward about twenty-five degrees, was sold without explanation.

It is always the policy of the C. L. Best Gas Traction Company to build a thing as it should be built. The great care with which each part of the Tracklayer is made and the manner in which the whole structure is designed is plainly to be seen in the pictures above. If you have a catalogue of other tractors at hand we suggest you compare the two motors as we call your attention to certain points.

Notice the clevise of the upper ends of the valve rods. Other mo-

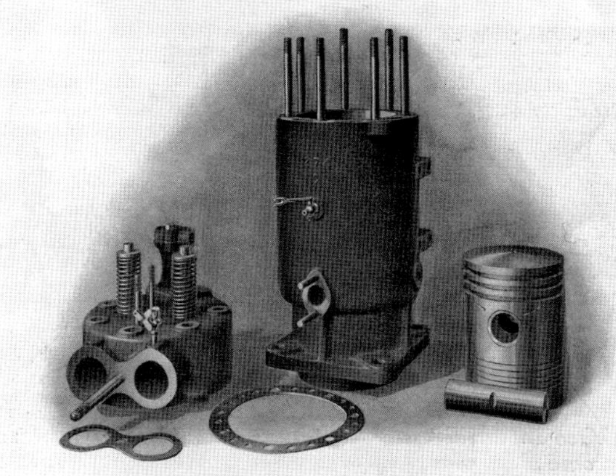

Removable cylinder heads and accessible valves are important in a tractor motor.

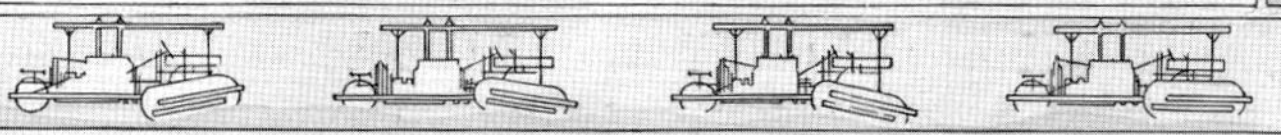

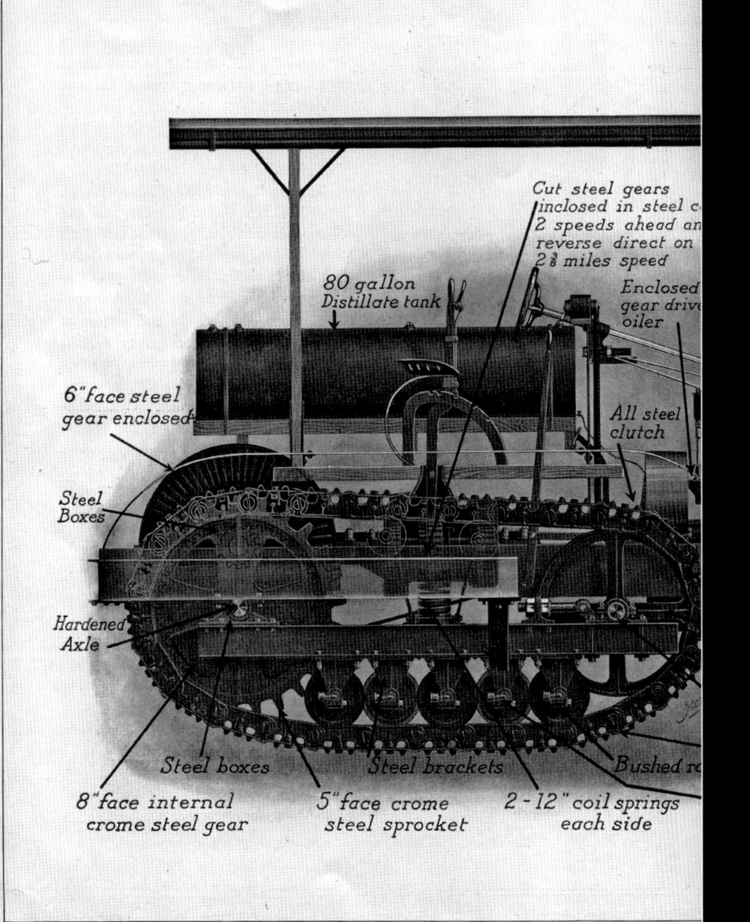
Cut steel gears
inclosed in steel c
2 speeds ahead an
reverse direct on
2⅜ miles speed
80 gallon
Distillate tank
Enclosed
gear drive
oiler
6" face steel
gear enclosed
All steel
clutch
Steel
Boxes
Hardened
Axle
Steel boxes
Steel brackets
Bushed r
8" face internal
crome steel gear
5" face crome
steel sprocket
2 - 12" coil springs
each side

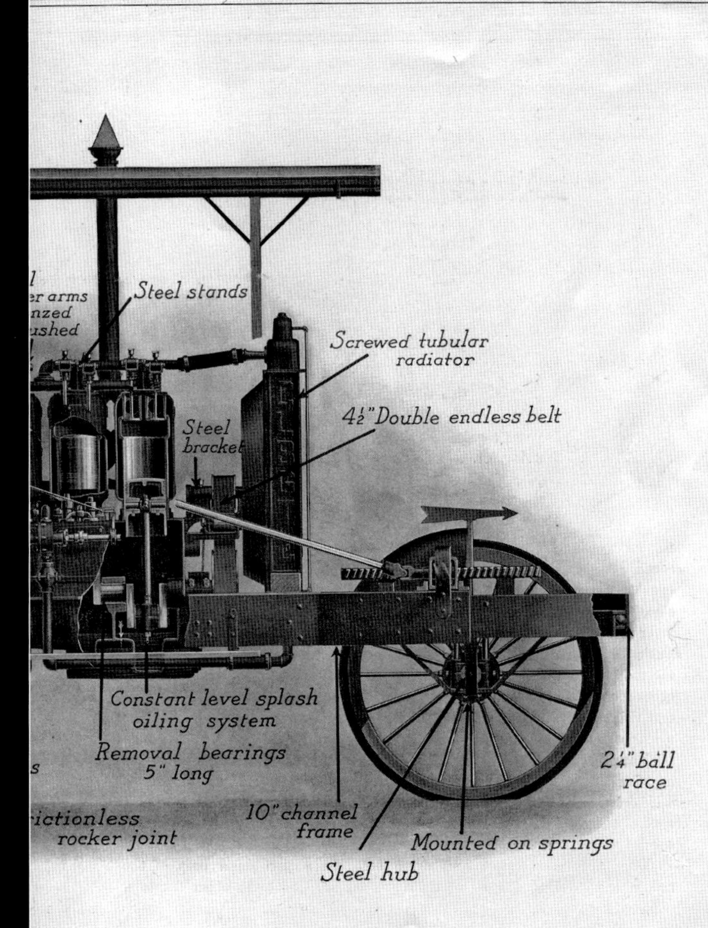
Steel stands
Screwed tubular
radiator
4½" Double endless belt
Steel
bracket
Constant level splash
oiling system
Removal bearings
5" long
rictionless
rocker joint
10" channel
frame
2¼" ball
race
Mounted on springs
Steel hub

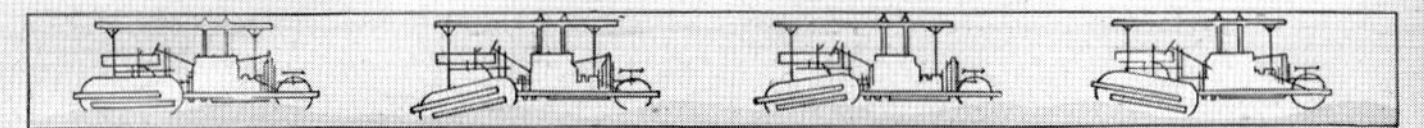

An example of Tracklayer accessibility. The piston comes out through the motor base.

tors are equipped with a rough ball and socket. The clevises cost much more, both to make and install, but they stand the wear and hard usage. See, also, the mechanical oiler which oils the cylinders. On the Best Tracklayer motor it is on the rear cylinder, carefully fitted into a niche especially made for it. Notice that it is gear driven; that this gear drive is a part of the motor itself, designed and installed when the motor was designed. On other motors this oiler will be found to be resting on a "makeshift" bracket and driven by a belt, the whole thing having been added long after the motor was originally built.

Your attention is also called to the magneto, water pump and governor, all of which are driven by a single shaft on the right side of the motor. This drive shaft and the placing of these fixtures plainly shows them to be a part of the original motor design. On other motors they will be found in several places and the water pump is driven by a belt.

The fan, fan bracket and drive belt are shown as a part of the motor and carefully placed in their present position. On other tractors,

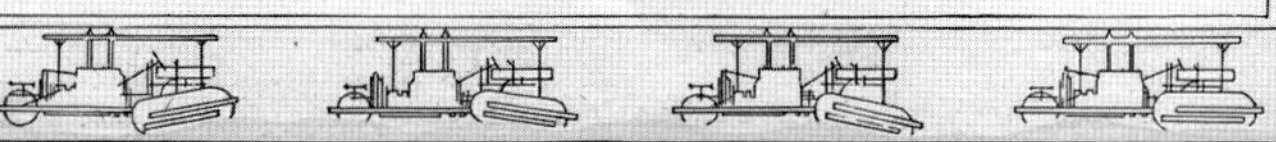

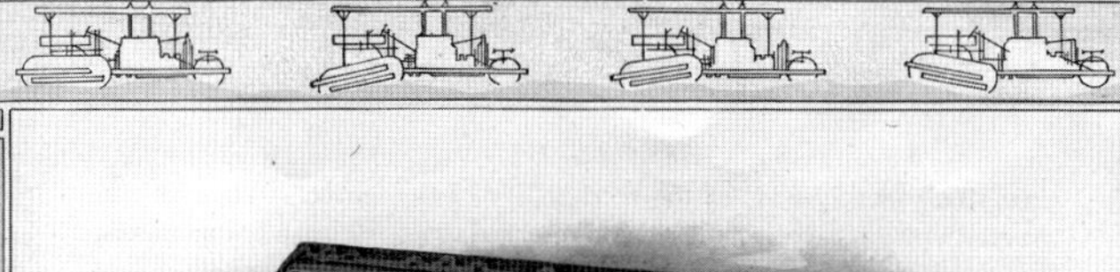

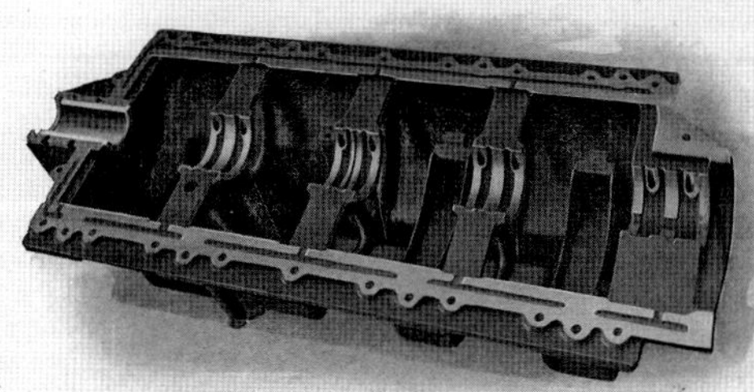

The bearings are all of ample size.

the fan is mounted in front of the radiator on the radiator platform. These fans revolve at the high speed of twelve to fifteen hundred revolutions each minute and unless solidly fixed to the motor the vibration is exceedingly strong. When placed on the radiator this vibration is sure to cause leakage in a short time. If springs are then placed under the radiator the vibration is only increased.

The fan bracket is adjustable. A single tension screw allows the belt to be tightened in an instant. The belt itself is four inches wide and double brass studded.

Best Tracklayer cylinders are cast separately with removable heads. The valves are in the heads and easily accessible at all times.

It is interesting to know that each piston in the Tracklayer motor may be removed from the cylinder and withdrawn through the side of the crank-

The fly-wheel is bolted to the flange on the left, not keyed.

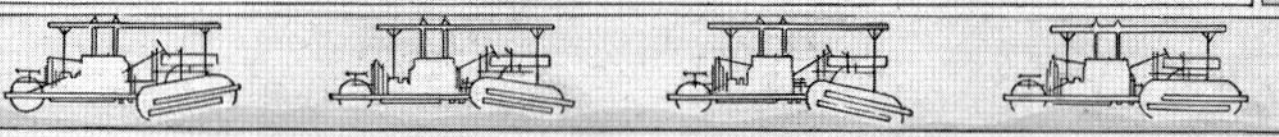

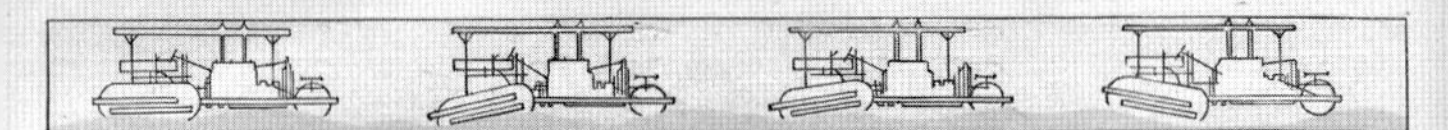

case. This greatly adds to the accessibility of the motor. When so constructed a piston may be removed and returned to its place in a few moments. It is unnecessary to remove the cylinder.

We are showing a picture of the crank-shaft used in the Tracklayer motor. We wish to call your attention to the flange on the left side, to which the fly-wheel is bolted. When this flange is missing, the fly-wheel is then keyed to the shaft. Only when the fly-wheel is bolted to a flange will it run absolutely true. A key forces it slightly out of line. It cannot be seen by the naked eye, yet a slight quiver is continually running through the shaft, finally causing crystallization and breakage.

The bearings in the crank-case are five inches long. The one next to the fly-wheel is nearly ten inches long. The front bearing is six and one-half inches. Since the Tracklayer motor was designed as a whole, the bearings and shaft were carefully drawn in size to conform to bore and stroke and rated horsepower. It is often found that motors which were originally designed for smaller cylinders and lower horsepower have since been increased in bore and stroke without increasing other dimensions.

The rocker arms on the Best Tracklayer motor are steel and bronze-bushed. The rocker arm stands are steel. The push rods are steel, hardened at the ends. Petticoats cover up the tappets, keeping the dirt out and the oil and grease in.

We rate our horsepower conservatively. Our motor has one inch longer stroke and one-quarter inch larger bore than other motors rated at the same horsepower, as you will find by comparison.

Notice the careful machining on these parts.

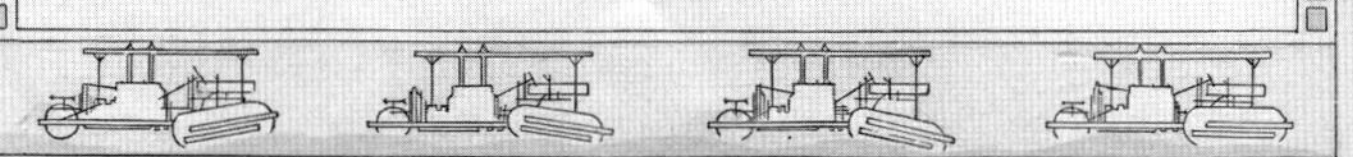

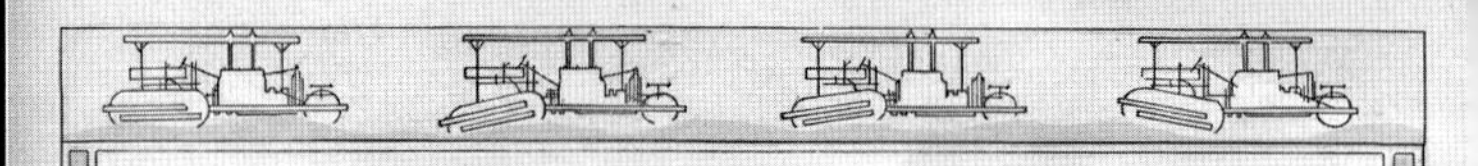

TRACK CONSTRUCTION

Naturally the most important and vital point in the construction of a track engine is the track and the manner in which it is built. If the track is not rightly constructed it is an unending source of trouble.

There never was a track construction as successful as that of the Best Tracklayer. The Best Tracklayer track alone, with none of the other superior points we will mention, would be sufficient to mark the engine as a wonderful improvement.

Its two distinguishing features are the track oscillation and the rocker joint in the links. The oscillating track eliminates the rigid front idler shaft attached to the frame of the tractor. This has two distinct advantages. It absolutely prevents the front wheel from rising in the air when the forward end of the track strikes a hummock or ditch. When the front of the track is rigidly attached to a solid axle which in turn is fastened to the frame, the front of the tracks, upon encountering a slight rise in the ground, naturally force the frame to rise also. When this occurs the front end of the tractor, carrying a 5,000-pound motor, immediately jumps up in the air, comes down with a mighty thud, jars the whole machine, racks the

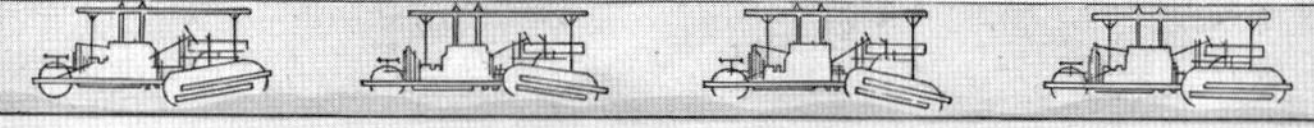

Track shoes, track links and the rocker joint.

motor and strains the frame. For instance, the fly-wheel on a 75 horse-power motor weighs about 600 pounds. It is attached to the end of the crank-shaft. A shock such as we have described would undoubtedly strain the shaft in a serious manner.

The front wheel of the Best Tracklayer always stays on the ground. When the front of the track strikes a hummock, the track oscillates upward. It is not solidly attached to the frame. The pictures in the border around this page illustrate this action.

There is another and equally important reason for the oscillating track. In plowing, a tractor always turns to the left. This brings the whole strain on the right-hand track, which as a consequence wears more than the other. Thus it is necessary to tighten the right-hand side oftener than the left. When the front end of the tracks are carried by a solid shaft attached to the frame it will soon be noticed that the sides of the tracks are wearing rapidly, more so than when the tractor was new. Investigation will show that the left front idler is probably three inches ahead of the right one, due to the more frequent tightening. Because the tracks are solidly attached to the frame, they are running, therefore, off at an angle to the frame and pro-

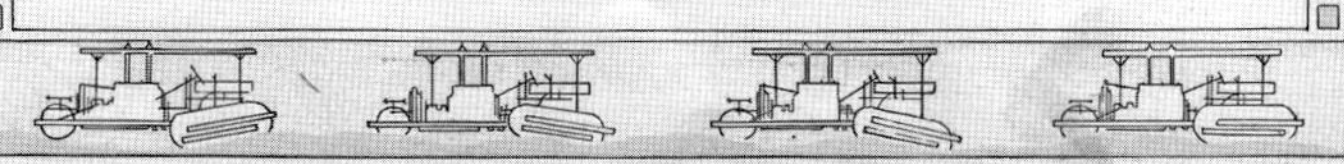

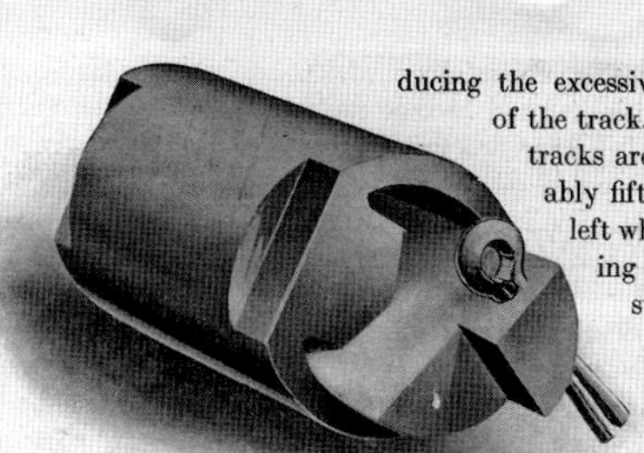

The rocker joint. The pin rocks from one flat side to another. There can be no friction.

ducing the excessive wear on the sides of the track. In other words, the tracks are often pointing probably fifteen degrees off to the left while the frame is pointing straight ahead. The steering wheel tries to go one way and the tracks another.

The Best Tracklayer tracks, having no rigid front idler shaft, are independent of the frame. This makes it possible to tighten either track as often as necessary without causing trouble. In fact, by the use of the Best Tracklayer type of construction it would be possible to successfully operate with one track six feet and the other ten feet long if such a freak were desirable.

Thus it is possible for the Best Tracklayer oscillating tracks to travel over checks or depressions without the front wheel rising from the ground, without strain of any kind on the frame, motor or other parts and, in turning, without excessive wear on the tracks. Common sense will also show that oscillating tracks prevent all side twisting and straining which occurs with the rigid tracks when one side mounts a hillock while the others travel on the level.

This brings us to the second prominent feature of the Best Tracklayer track, the rocker joint. We were forced to evolve the rocker joint in the track links because we soon discov-

Rollers, drive sprocket and extended track.

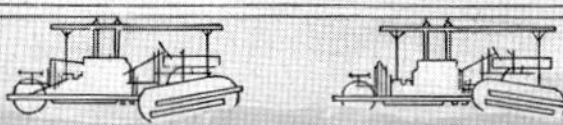
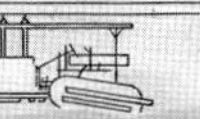
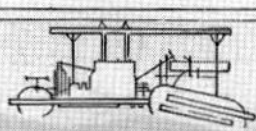
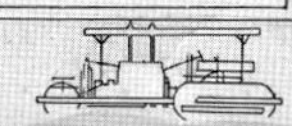

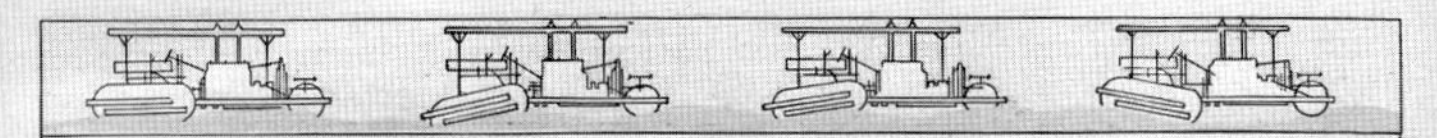

ered that tracks cannot be successfully oiled. Oil holes in the pins cannot be used. They fill with dirt and become useless. It is then necessary to drown the tracks in oil if they are to last any length of time. In loose soil or in sand this flood of oil soon produces a thick paste almost as tough as asphalt and affording no lubrication. It acts, rather, in the opposite way. Such a sticky mass, filled with sand soon cuts the pins and links to pieces.

We therefore designed our rocker joint. It positively requires no oil. It couldn't be oiled if one were to try. It is frictionless. When the tracks bend while passing over the sprockets the joint simply rocks from one flat side to another. There is no friction and no outer sleeve to cause friction. It is this exclusive feature which allows us to guarantee the wear of our track for one year, as the written guarantee on another page shows. No other manufacturer in the market could afford a similar guarantee.

Automatic steering device, operated by the motor.

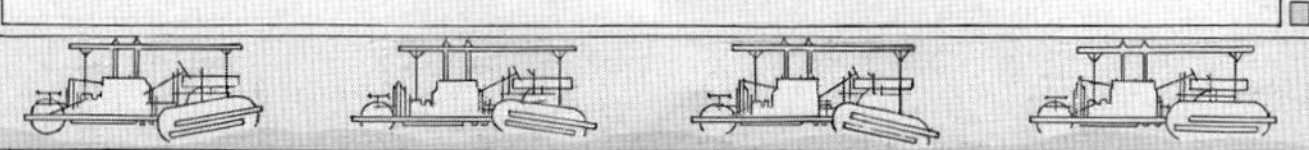

TRANSMISSION

In a tractor as carefully built as the Best Tracklayer we should not, under any circumstances, consider using anything but geared transmission. All the well-known and reputable tractors in the United States, both round-wheel and otherwise, use the gear drive. Manufacturers who have been building machines for a score of years and whose construction has become standard all agree on this one point: the chain drive should never be used for heavy-duty work. If you will call to mind the many styles of machines with which you are more or less acquainted you will easily remember their geared construction. Allow no one to tell you that chains have the slightest excuse for existence on a tractor.

The Best Tracklayer transmission has two speeds ahead and one reverse. The gears are cut steel and as the illustrations show, are enclosed in dust-proof cases and revolve in oil constantly. Unlike some other manufacturers, no extra charge is made for the two speeds ahead. This is a part of the regular construction. Its advantages are many. There are frequent

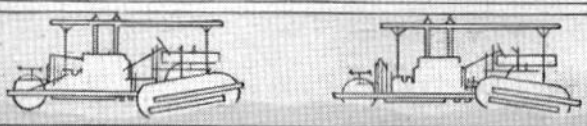
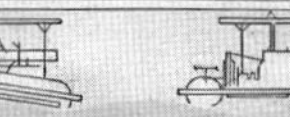
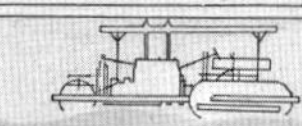

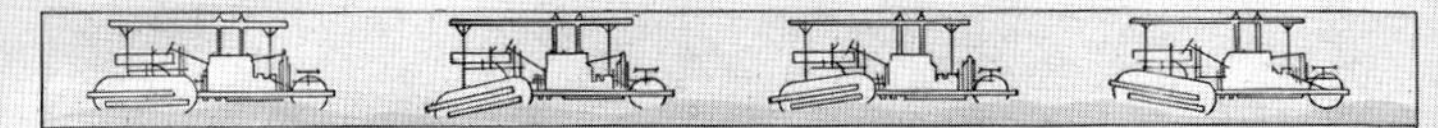

Differential and drive sprockets.

occasions when steep grades or difficult operating conditions are made easier by the use of a low speed.

The countershaft on the Best Tracklayer drives two internal gears, one on each side, each with an eight-inch face and made of chrome steel. Cast integral with each internal gear is the track sprocket, which drives the tracks. These are shown above. They have a five-inch face and, as we have said above, are made of chrome steel. By turning to the large picture on pages 18 and 19, the transmission construction will be entirely clear. Compare this with chain sprockets with about a two-inch face and a chain from countershaft to drive shaft.

Specifications 75 h-p. Tracklayer

MOTOR—Four-cylinder, four-cycle, water cooled. Cylinders cast separate, heads removable, valves in heads. Steel rocker arms and stands, bronze bushed. All parts easy to get at and easily removed.

BORE AND STROKE—7¾-inch bore, 9-inch stroke. Mechanically operated overhead valves, spiral cut steel timing gears, wide face. Crank shaft of special hammered steel ground to 1-1000th of an inch. Flange for fly-wheel forged solid; all bearings generous length, removable, and of white bronze. Interchangeable parts throughout.

POWER—75 brake horse-power at 450 revolutions per minute; 82 brake horse-power at 500 revolutions per minute. Pistons removable through inspection plates.

TRANSMISSION—Two speeds ahead, one reverse; all steel cut gears, wide face. Low speed 1½ miles per hour. High speed 2¼ miles per hour. "Direct" reverse 1⅝ miles per hour. No reduction gears on plowing speed. Drive shaft to a 6-inch face steel differential gear with two side clutches for short turning; 8-inch face steel bull pinion to 8-inch face steel internal gear. 5-inch face drive sprocket cast integral with internal gear. All gears enclosed and run in oil.

IGNITION—High Tension Dual Magneto, dirt-proof.

CARBURETOR—Schebler, 2 inches. Model E.

GOVERNOR—"C. L. B." make, enclosed and runs in oil, but one moving lever exposed. Simple and serviceable, and governs closely.

WATER CIRCULATING PUMP—Rotary gear, driven by enclosed cut steel gear.

CIRCULATING OIL PUMP—Rotary gear driven by enclosed cut steel gear.

BEARINGS—Combined length of motor bearings, 51 inches. Removable saddles and die cast.

COOLING—Vertical tube radiator with screwed tubes.

FAN—All steel in steel adjustable bracket; ball thrust collar. Driven with 4-inch double belt.

PISTON PIN—Connecting rod, clamped to pin with clamp bolt, passing through side of pin. Pin journals in piston, which is manganese bronze bushed, affording a long bearing and eliminating a piston pin set screw.

TRANSMISSION CASE—Steel transmission case, 10½-inch and 8-inch bearings.

WEIGHT—Fully equipped, 24,800 lbs.

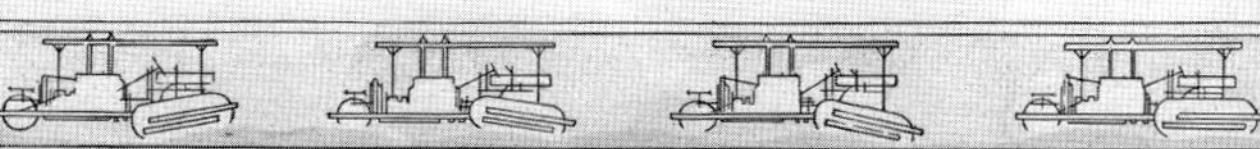

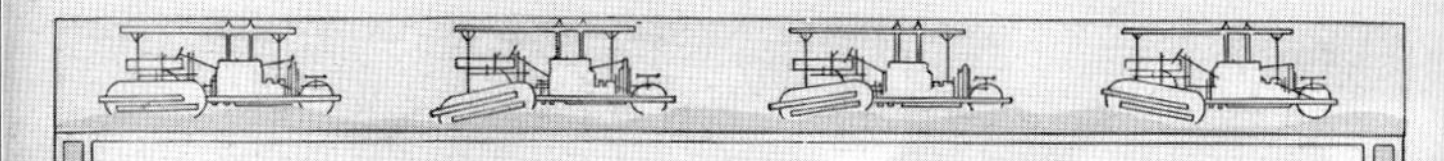

30 H-P ORCHARD TRACKLAYER

Four essentials must be embodied in the successful orchard tractor: short turning radius, light weight, compactness, and ample power. Considerable difficulty has been experienced in attempting their combination. Ample power seemed to mean excessive weight; excessive weight made short turns difficult.

But the Best Orchard Tracklayer has approached closer to the needed requirements than any similar machine. It will turn squarely into one row from another. The weight is 7,600 pounds equipped, fully 2,000 pounds or one-third lighter than similar machines. The length is but 11 feet 8 inches compared to 14 or 15 feet. The height is 59 inches and the width 65 inches, the former at least eleven inches lower and the latter five inches narrower than other tracklaying tractors. And the four cylinders, 5¼ x 6¼, produce thirty-brake or sixteen drawbar horsepower. Our cylinders are a quarter-inch longer in stroke than similar machines rated at equal horsepower, and we guarantee sixteen drawbar horsepower compared to the fourteen of the nearest competitor.

We have prepared an illustrated booklet containing a complete description of this small tractor and its construction. You should write for it.

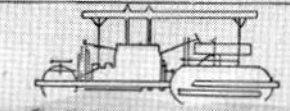

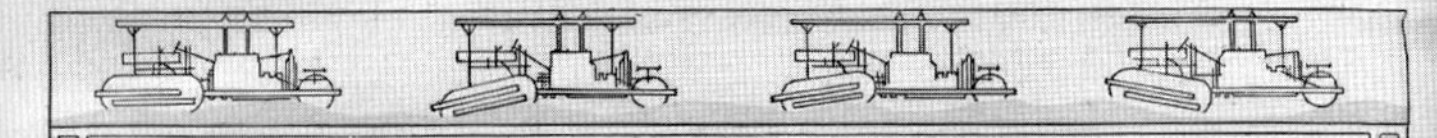

Short turning, light weight, compactness and ample power are needed in the orchard tractor. Here you find all of these essentials fully realized.

Complete Specifications of 30 h-p. Tracklayer

MOTOR—Four-cylinder, four-cycle, water cooled. Cylinders cast separate with valves in head. Heads removable.

BORE—5¼-inch.

STROKE—6¼-inch.

RATING—30-B. H. P. continuously at 600 R. P. M.

IGNITION—Dry cell battery with Bosch Magneto and Coil.

COOLING—Centrifugal pump, vertical copper tube radiator; 2½-inch double fan belt—easily adjusted.

LUBRICATION—Constant level splash system with gear oil pump. Oil and grease cups placed where needed.

GOVERNOR—C. L. B. Automatic Governor enclosed and run in oil. But one operating lever exposed.

CRANK SHAFT—40-carbon open hearth machine steel drop forged 2½-inch diameter; five crank shaft bearings. Total length, 21¾ inches.

TRANSMISSION—Drive, forged bevel pinion to 2 bevel gears with reverse and direct drive expanding band clutches, one in each bevel gear. High and low spur, wide face steel pinions on same shaft, which drive two large spur gears with independent side clutches of the expanding band type in each gear. Drive sprocket in on this same shaft and drives direct on to the track, doing away with the third motion drive with chains or gears. All steel cut gears enclosed and run in oil.

DRIVE SHAFTS—First motion shaft, 2 11-16 inches diameter, nickel steel; Hess bright ball-bearing mounting. Second motion shaft, steel boxes, 3-inch diameter, two bearing. Total length 18 inches.

SPEEDS—Two forward and reverse on either speed. Two miles per hour on low and 2¾ miles on high.

CARRYING WHEELS—Two steel carrying wheels on each side carry the weight, also ball race roller, which supports the weight and takes the side thrust at the center of the track.

TRACK—Consists of C. L. B. rocker joint frictionless chain. Spool, which sprocket drives, has a diameter of 2 inches with a 3-inch face. Rocker joint enclosed. Manganese spool. Hardened pin.

TRACK SHOES—Drop forged plow steel over-lapping. Standard sizes, 14-inch and 16-inch. Can be fitted with smaller shoes to order.

FRAME—Built up frame of angles and plates with a depth of 12 inches at center tapering off to 6 inches at each end and tied together with cross channels, supported by a gusset plate over the entire bottom, doing away with rods and braces. Joints hot riveted.

FRONT WHEEL—Cast steel, 26-inch diameter, 10-inch face, mounted on springs with a vertical travel of 6 inches, which forms itself to any ground; self-cleaning scrapers, with ball-bearing wheel circle.

STEERING GEAR—Steering column is mounted vertically in front of driver with control levers, one on each side.

TANK CAPACITY—Gasoline, 4 gallons. Distillate, 25 gallons.

PRINCIPAL DIMENSIONS—Length over all, end to end of frame, 11 feet 8 inches. Seat projects over 2 feet at rear of frame. Width over all, 5 feet 5 inches, with 14-inch shoe. Width, center to center of track, 4 feet 3 inches. Height over all, 4 feet 11 inches. From center of front wheel to center of track, 7 feet.

WEIGHT—Weight, fully equipped, 7,600 lbs.

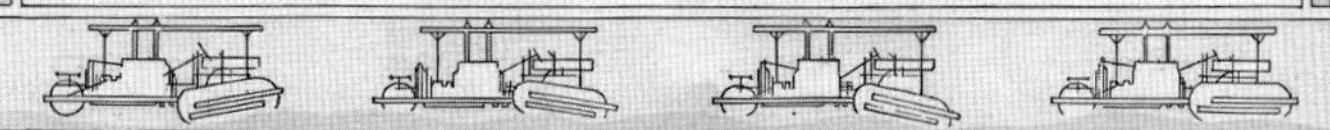

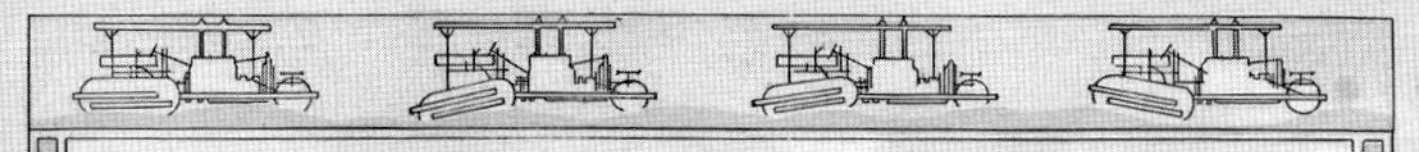

Specifications 75 h-p. Best Round-Wheel Tractor

We still build round-wheel tractors for those who want them for use in favorable conditions. And no better round-wheel tractors are built anywhere.

We have a complete catalogue describing them. You may have one, if interested. In the meantime, these are the specifications:

MOTOR	
Number of Cylinders	4
Bore	7¾
Stroke	9
POWER (H. P.)	
Rated	75
Actual brake—	
at 450 R. P. M.	75
at 500 R. P. M.	82
Drawbar power—	
1¾ miles per hour	50
2⅓ miles per hour	40
3¾ miles per hour	20
3 speeds ahead and reverse	
WHEELS—MAIN	
Diameter	90
Width	24–36
TILLER WHEEL	
Diameter	50″
Width	24″

BEARINGS	
Length of main bearings	29
Length of crank pin bearings	5
Length of crank bearings	5
Combined length of crank and wrist pin bearings	66
Number of crank bearings	5
DIMENSIONS	
Length over all	21′ 5″
Width over all	10′ 7″
Width between wheels	55″
To top of canopy	10′ 4″
To top of wheel housing	8′
TANK CAPACITY	
Water	60
Distillate	80
Gasoline	5
Lubricating oil	13
CLEARANCE	30
WEIGHT, lbs.	24,000

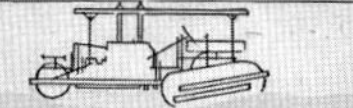
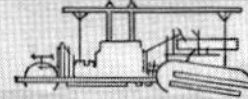
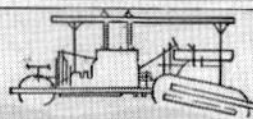
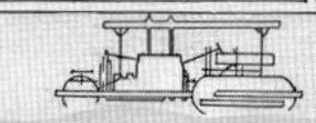

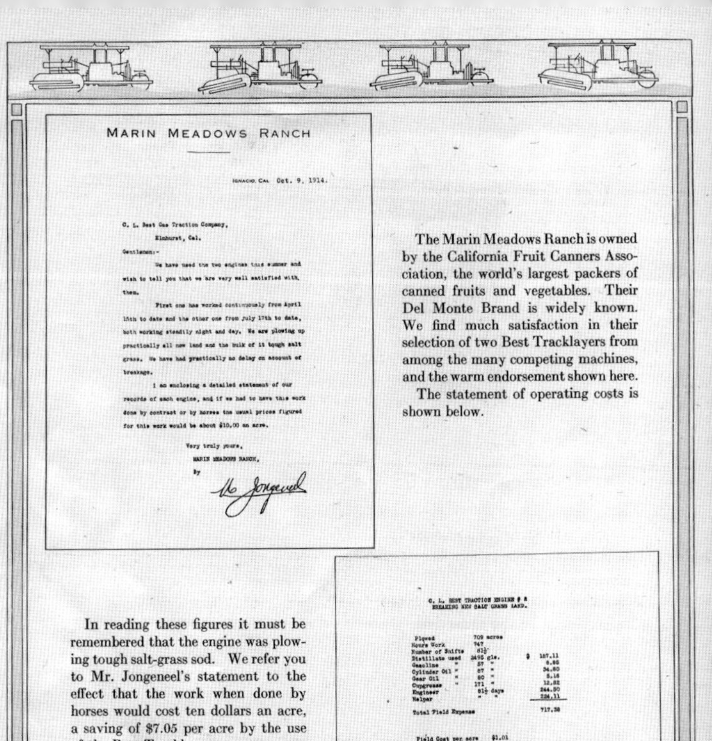

MARIN MEADOWS RANCH

IGNACIO, CAL. Oct. 9, 1914.

C. L. Best Gas Traction Company,

Elmhurst, Cal.

Gentlemen:-

We have used the two engines this summer and wish to tell you that we are very well satisfied with them.

First one has worked continuously from April 15th to date and the other one from July 17th to date, both working steadily night and day. We are plowing up practically all new land and the bulk of it tough salt grass. We have had practically no delay on account of breakage.

I am enclosing a detailed statement of our records of each engine, and if we had to have this work done by contrast or by horses the usual prices figured for this work would be about $10.00 an acre.

Very truly yours,

MARIN MEADOWS RANCH,

By

M. Jongeneel

The Marin Meadows Ranch is owned by the California Fruit Canners Association, the world's largest packers of canned fruits and vegetables. Their Del Monte Brand is widely known. We find much satisfaction in their selection of two Best Tracklayers from among the many competing machines, and the warm endorsement shown here.

The statement of operating costs is shown below.

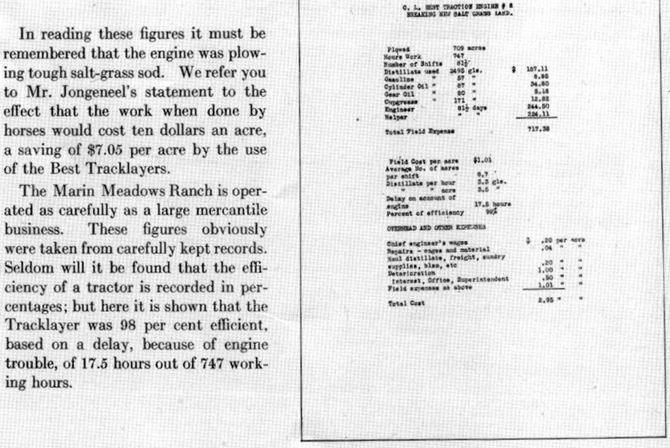

In reading these figures it must be remembered that the engine was plowing tough salt-grass sod. We refer you to Mr. Jongeneel's statement to the effect that the work when done by horses would cost ten dollars an acre, a saving of $7.05 per acre by the use of the Best Tracklayers.

The Marin Meadows Ranch is operated as carefully as a large mercantile business. These figures obviously were taken from carefully kept records. Seldom will it be found that the efficiency of a tractor is recorded in percentages; but here it is shown that the Tracklayer was 98 per cent efficient, based on a delay, because of engine trouble, of 17.5 hours out of 747 working hours.

C. L. BEST TRACTION ENGINE # 2
BREAKING NEW SALT GRASS LAND.

Plowed	709 acres	
Hours Work	747	
Number of Shifts	81½	
Distillate used	2495 gls.	$ 187.11
Gasoline "	57 "	8.85
Cylinder Oil "	87 "	34.80
Gear Oil "	80 "	5.16
Cupgrease "	171 "	12.82
Engineer	81½ days	244.50
Helper	" "	224.11
Total Field Expense		717.38

Field Cost per acre	$1.01
Average No. of acres per shift	8.7
Distillate per hour	3.3 gls.
" " acre	3.5 "
Delay on account of engine	17.5 hours
Percent of efficiency	98%

OVERHEAD AND OTHER EXPENSES

Chief engineer's wages	$ .20 per acre
Repairs - wages and material	.04 " "
Haul distillate, freight, sundry supplies, blsm, etc	.20 " "
Deterioration	1.00 " "
Interest, Office, Superintendent	.50 " "
Field expenses as above	1.01 " "
Total Cost	2.95 " "

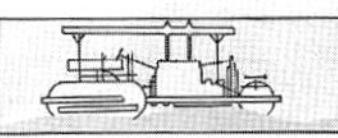
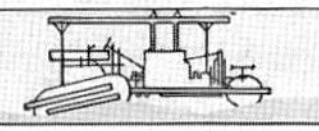
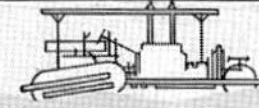
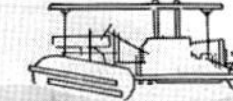

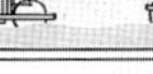
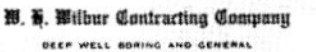

W. H. Wilbur Contracting Company
DEEP WELL BORING AND GENERAL CONSTRUCTION WORK

Tulare, Cal., Oct. 18th 1914.

C. L. Best Gas Traction Co.

Oakland, Calif.

Gentlemen:-

I am in receipt of yours of the 6th asking for my experience with the 70 H. P. Gas tractor purchased from you this season.

I put this engine to work on June 15th pulling a 37 foot cut steam harvester changed into a gasoline rig and ran until Sept. 11. We ran every working day except July 3rd and 4th, making 75 days continuous run. During this time I have had no engine trouble nor no repairs whatever.

From July 24th until Aug. 27, which covered the heaviest part of my harvest, my fuel consumption was 47 gallons of distillate for each 10 hours and 1.64 gals of oil for each 10 hours run.

I have been operating steam and gasoline tractors in this section for the past seven years and I am pleased to say that the C. L. Best 70 H. P. gas tractor fills every want for a heavy duty traction engine.

Very truly yours,

W H Wilbur

W. H. Wilbur is a very well-known California farmer. Not only has he been using tractors for seven years—he has used probably an equal number of different tractors. When he indorses the Best Tracklayer he does so from an intimate knowledge of all the machines used in this State, both tracklaying and otherwise.

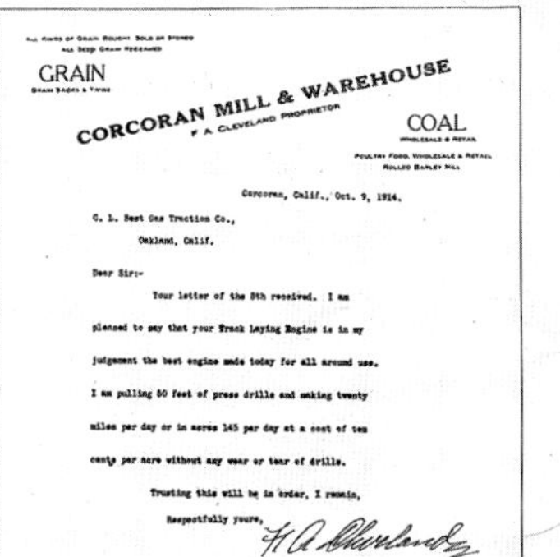

GRAIN
GRAIN SACKS & TWINE

CORCORAN MILL & WAREHOUSE
F. A. CLEVELAND PROPRIETOR

COAL
WHOLESALE & RETAIL

POULTRY FOOD, WHOLESALE & RETAIL
ROLLED BARLEY MILL

Corcoran, Calif., Oct. 9, 1914.

C. L. Best Gas Traction Co.,

Oakland, Calif.

Dear Sir:-

Your letter of the 8th received. I am pleased to say that your Track Laying Engine is in my judgement the best engine made today for all around use. I am pulling 60 feet of press drills and making twenty miles per day or in acres 145 per day at a cost of ten cents per acre without any wear or tear of drills.

Trusting this will be in order, I remain,

Respectfully yours,

F A Cleveland

Everybody near Corcoran is acquainted with Mr. Cleveland. To have this expression of satisfaction from him is gratifying indeed. Don't hesitate to write to him, if you care to have a more detailed opinion.

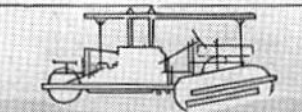
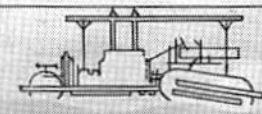
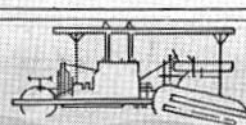
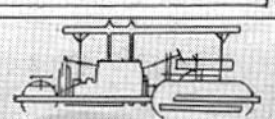

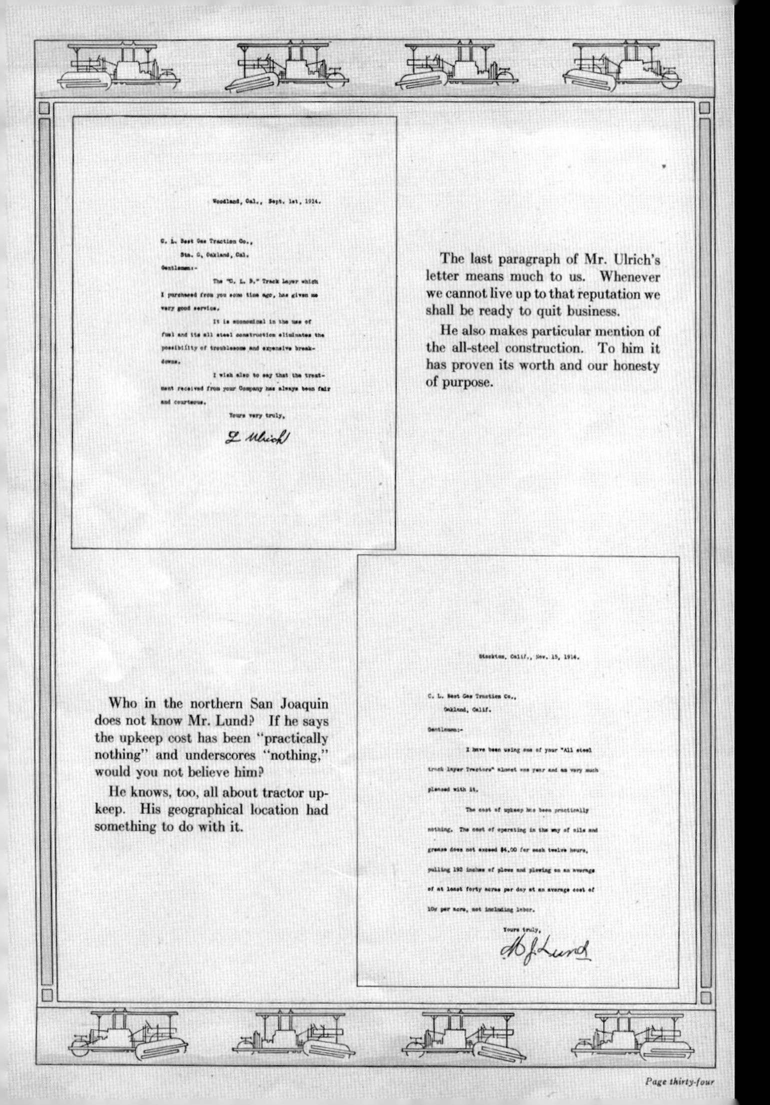

Woodland, Cal., Sept. 1st, 1914.

C. L. Best Gas Traction Co.,
Sta. G, Oakland, Cal.

Gentlemen:-

The "C. L. B." Track Layer which I purchased from you some time ago, has given me very good service.

It is economical in the use of fuel and its all steel construction eliminates the possibility of troublesome and expensive breakdowns.

I wish also to say that the treatment received from your Company has always been fair and courteous.

Yours very truly,

L Ulrich

The last paragraph of Mr. Ulrich's letter means much to us. Whenever we cannot live up to that reputation we shall be ready to quit business.

He also makes particular mention of the all-steel construction. To him it has proven its worth and our honesty of purpose.

Who in the northern San Joaquin does not know Mr. Lund? If he says the upkeep cost has been "practically nothing" and underscores "nothing," would you not believe him?

He knows, too, all about tractor upkeep. His geographical location had something to do with it.

Stockton, Calif., Nov. 15, 1914.

C. L. Best Gas Traction Co.,
Oakland, Calif.

Gentlemen:-

I have been using one of your "All steel truck layer Tractors" almost one year and am very much pleased with it.

The cost of upkeep has been practically nothing. The cost of operating in the way of oils and grease does not exceed $4.00 for each twelve hours, pulling 192 inches of plows and plowing on an average of at least forty acres per day at an average cost of 10¢ per acre, not including labor.

Yours truly,

H. J. Lund

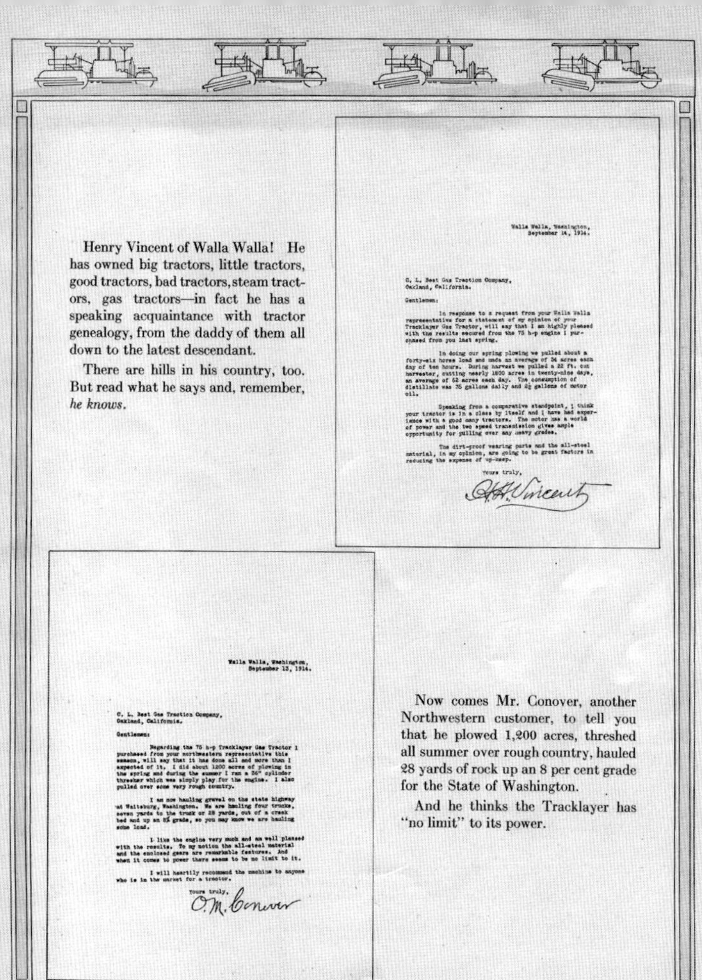

Henry Vincent of Walla Walla! He has owned big tractors, little tractors, good tractors, bad tractors, steam tractors, gas tractors—in fact he has a speaking acquaintance with tractor genealogy, from the daddy of them all down to the latest descendant.

There are hills in his country, too. But read what he says and, remember, *he knows.*

Walla Walla, Washington,
September 14, 1914.

C. L. Best Gas Traction Company,
Oakland, California.

Gentlemen:

In response to a request from your Walla Walla representative for a statement of my opinion of your Tracklayer Gas Tractor, will say that I am highly pleased with the results secured from the 75 h-p engine I purchased from you last spring.

In doing our spring plowing we pulled about a forty-six horse load and made an average of 34 acres each day of ten hours. During harvest we pulled a 22 ft. cut harvester, cutting nearly 1800 acres in twenty-nine days, an average of 62 acres each day. The consumption of distillate was 35 gallons daily and 2½ gallons of motor oil.

Speaking from a comparative standpoint, I think your tractor is in a class by itself and I have had experience with a good many tractors. The motor has a world of power and the two speed transmission gives ample opportunity for pulling over any heavy grades.

The dirt-proof wearing parts and the all-steel material, in my opinion, are going to be great factors in reducing the expense of up-keep.

Yours truly,

H. H. Vincent

Walla Walla, Washington,
September 13, 1914.

C. L. Best Gas Traction Company,
Oakland, California.

Gentlemen:

Regarding the 75 h-p Tracklayer Gas Tractor I purchased from your northwestern representative this season, will say that it has done all and more than I expected of it. I did about 1200 acres of plowing in the spring and during the summer I ran a 36" cylinder thresher which was simply play for the engine. I also pulled over some very rough country.

I am now hauling gravel on the state highway at Waitsburg, Washington. We are hauling four trucks, seven yards to the truck or 28 yards, out of a creek bed and up an 8% grade, so you may know we are hauling some load.

I like the engine very much and am well pleased with the results. To my notion the all-steel material and the enclosed gears are remarkable features. And when it comes to power there seems to be no limit to it.

I will heartily recommend the machine to anyone who is in the market for a tractor.

Yours truly,

O. M. Conover

Now comes Mr. Conover, another Northwestern customer, to tell you that he plowed 1,200 acres, threshed all summer over rough country, hauled 28 yards of rock up an 8 per cent grade for the State of Washington.

And he thinks the Tracklayer has "no limit" to its power.

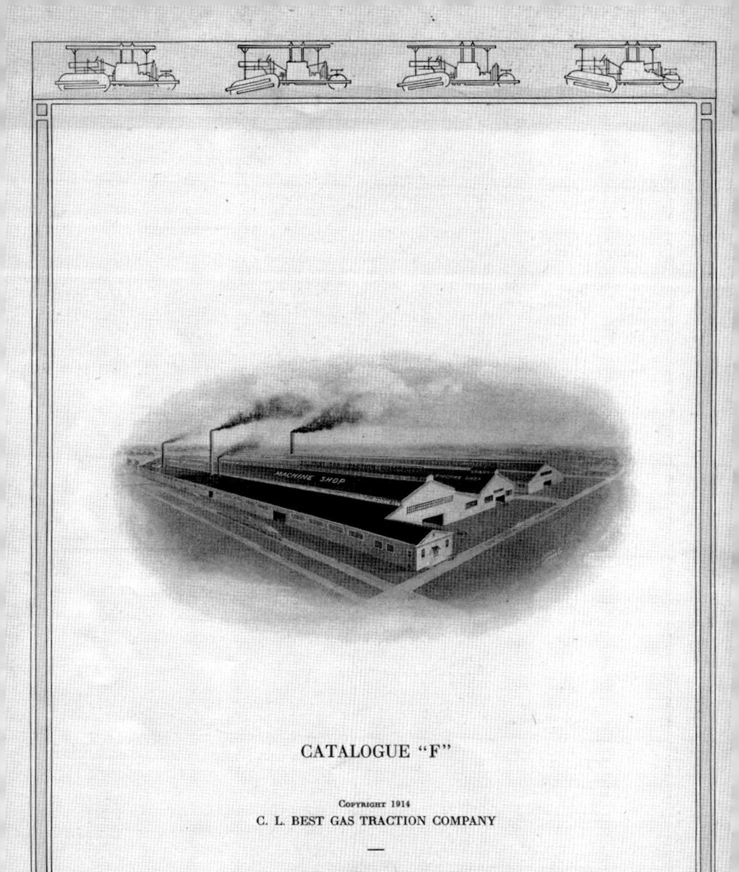

CATALOGUE "F"

—

Written by DOUGLAS BOSWELL
Designed by HONIG ADVERTISING COMPANY
Printed by THE BLAIR-MURDOCK CO.

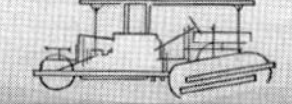
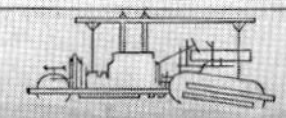
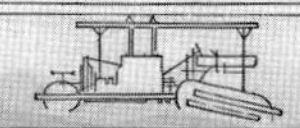
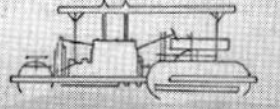

INDEX

Page numbers in *italics* refer to illustrations